SCIENCE AND SOCIOLOGY

Science and Sociology is from beginning to end an exploration of what this implies for the social sciences, and sociology in particular. The authors argue that over the last several decades, sociology has become less a science and more a quest for isolated assessments of situations, whether they come from demographic analyses, survey research, or ethnographic studies. Above all else, this book is an attempt to promote and advance scientific sociology, and we write at length specifying the how and why of this objective. With this objective in mind, the question becomes: What would a scientific sociology look like?

Sheldon Ekland-Olson joined The University of Texas at Austin after completing his graduate work at the University of Washington in Seattle and Yale Law School. He is currently the Bernard and Audre Rapoport Centennial Professor of Liberal Arts and serving as the Director of the School of Human Ecology. For five years he served as Dean of the College of Liberal Arts and then for eight years as Executive Vice President and Provost of the university. He has authored or co-authored several books and numerous articles on criminal justice, prison reform, and capital punishment. Widely recognized for his commitment to teaching undergraduates, he is the recipient of numerous teaching awards. His current interests are reflected in the book, *Who Lives, Who Dies, Who Decides?*, an exploration of how communities have gone about justifying the violation of universally held moral imperatives.

Jack P. Gibbs graduated with his PhD from University of Oregon in 1957. In addition to holding a faculty position at Vanderbilt, Gibbs also taught at the University of California, Berkeley, the University of Texas, Austin, Washington State University, and the University of Arizona. At Vanderbilt, he was the Centennial Professor of Sociology, as well as chairing the Department of Sociology and Anthropology. Dr. Gibbs has well over 170 publications, including scholarly articles in referred academic journals and books. He has received numerous scholarly honors, including a Guggenheim Fellowship, election as a fellow of the American Society of Criminology, and recipient of the 1983 Edwin Sutherland Award from the American Society of Criminology. His research areas included suicide, status integration, research methods, urbanization and technology, sociology of law, criminal deterrence, conceptions of deviant behavior, capital punishment, social control, and terrorism. He has authored these books: *Sociological Theory Construction; Crime, Punishment, and Deterrence; Norms, Deviance, and Social Control; Control: Sociology's Central Notion; A Theory About Control*; and *Colossal Control Failures*.

SCIENCE AND SOCIOLOGY

Predictive Power is the Name of the Game

Sheldon Ekland-Olson and Jack P. Gibbs

Routledge
Taylor & Francis Group
NEW YORK AND LONDON

First published 2018
by Routledge
711 Third Avenue, New York, NY 10017

and by Routledge
2 Park Square, Milton Park, Abingdon, Oxon OX14 4RN

Routledge is an imprint of the Taylor & Francis Group, an informa business

Library of Congress Cataloging-in-Publication Data
Names: Ekland-Olson, Sheldon, 1944- author. | Gibbs, Jack P., author.
Title: Science and sociology : predictive power is the name of the game /
Sheldon Ekland-Olson and Jack P. Gibbs.
Description: New York, NY : Routledge, [2017] | Includes bibliographical references.
Identifiers: LCCN 2017000130 | ISBN 9781138047839 (hardback : alk. paper) |
ISBN 9781138047846 (pbk. : alk. paper) | ISBN 9781315170565 (e-book)
Subjects: LCSH: Sociology--Philosophy. | Sociology--Methodology. | Science--Social
aspects.
Classification: LCC HM511 .E44 2017 | DDC 301.01--dc23
LC record available at https://lccn.loc.gov/2017000130

ISBN: 978-1-138-04783-9 (hbk)
ISBN: 978-1-138-04784-6 (pbk)
ISBN: 978-1-315-17056-5 (ebk)

Typeset in Bembo
by Taylor & Francis Books

To the memory of: George C. Homans and Robert K. Merton

CONTENTS

ILLUSTRATIONS

Figures

Table

PREFACE

This book grew out of a belief, held by both authors, that the major objective for science, including sociology in particular, is the search for predictable uniformities. In the case of Gibbs, this belief is manifest in a decades-long search for uniformities in suicide rates across a variety of populations and over time (e.g., Gibbs and Martin 1964 and Gibbs 2001). For Ekland-Olson it has been a more recent matter of uncovering the predictable uniformities in how communities justify the violation of deeply important moral imperatives regulating the taking of life and tolerance of suffering (e.g., Ekland-Olson 2012).

Other extensive efforts can be found that seek to identify predictable uniformities in social life across time, place, and life circumstances (e.g., Black 1976 and 2011), but our argument is that over the last several decades sociology has become less a science and more a quest for isolated assessments of situations, be they demographic analysis, survey research, or ethnographic studies. More simply and perhaps cynically stated, the sociological enterprise has been focused on the promise of promoting social change and thereby making the world a better place. Ignoring the implied moral arrogance of this pursuit, the suggestion is that sociologists know how to change society. This is simply not true. At most, sociologists (including us) know very little about such matters. Knowledge to reshape a society in predictable ways must come in the form of theories, and for all practical purposes sociologists have no theories about promoting societal change, let alone tested or even testable theories.

Theories, you say, are *not* necessary to change a society? Nonsense! Any proposed policy or plan of action, with a well-defined sought after outcome, is based on a theory, be it explicit or implicit. Whether a policy/program will have the anticipated outcome depends on the related theory's predictive power, a subject treated and promoted at length in this book, especially as regards distinct dimensions of predictive power.

Above all else, this book is an attempt to advance scientific sociology, and we write at length specifying this objective. Our promotion of a science-based sociology is justified by this question: what is the alternative? If it is to be other than a source of situational assessments and the promotion of change, sociology must have a distinctive focus. In that connection, sociologists are not trained to be social workers. Nor can the primary reason for the profession be the investigation of social problems with a view to exposés. Journalists and novelists can and do play that role far more effectively. Consider John Steinbeck's *Grapes of*

Wrath and Harriet Beecher Stowe's *Uncle Tom's Cabin.* Both had a far greater impact on social change in the U.S. than all of the publications of C. Wright Mills, the model sociologist-as-ideologue (granted his admirable writing ability and keen eye for the important).

Our own ideology? We are left-leaning, some would say yellow-dog Democrats. But the idea that science has some inherent ideological orientation (liberal, progressive, socialist, conservative, or reactionary) is a myth. Consider the range of political leanings in sociology's pantheon, notably Comte, Durkheim, Marx, Pareto, Simmel, Spencer, and Weber. Surely the comparisons would undermine the argument that some particular ideological orientation is inevitable among sociologists.

Setting that argument aside, the question becomes: what would a scientific sociology look like? Predictive, testable assertions will form its distinguishing characteristics. We commence (Chapter 1) with the importance of inherent uncertainty in science and underscore the idea of assertions as distinguished from facts. We then progress through a series of arguments calling for testable theories and assessments of them based on their predictive power relative to that of contending theories. The related implicit suggestion is that sociologists cannot make progress by continually warming their hands over the bones of ancestors and perpetuating exegetical interpretations. Those ancestors did not play a major role in what we describe as "disastrous beliefs" in sociology (Chapter 8). Far from it; the current anti-science orientation among sociologists is a much more recent development.

The very notion of science may appear to eliminate any need for an extensive treatment of theory construction; but some sociologists evidently do not identify themselves as scientists, and many others have quite different conceptions of science, including "scientific theories." Chapter 3 makes it clear that divergent conceptions of science are not peculiar to sociologists. That recognition led us to set forth our conception of science and scientific theories at length in Chapters 4 and 5, including identification of seven dimensions of predictive power.

Our promotion of scientific sociology is also focused on a mode of formal theory construction (Chapters 6 and 7), one designed to promote systematic tests of theories. However, in Chapter 9 we concede that the search for uniformities and statements of them as propositions should take priority.

The belief that science deals with assertions, not facts known with absolute certainty, led us to commence the book with a treatment of assertions, the exclusive subject of Chapter 1. However, we never tire of emphasizing uncertainties in science, and in Chapter 2 we argue that the uncertainties are largely if not entirely a matter of the predictive implications of assertions. Positivism? Let us hope so.

Well, there you have it. We move from the inherent importance of assertions, through the assessment criterion of relative predictive power, to the construction of formal theories. We present these ideas in full recognition that what we advocate is admittedly more about a hoped-for future than a celebration of accomplishment.

To be clear, this book is an attempt to promote scientific sociology and a categorical rejection of sociology as the promotion of some ideology, be it Marxism, feminism, or environmentalism. Some might respond: how could your promotion of scientific sociology be anything other than the promotion of an ideology? Answering with a rhetorical question: is the explication of the rules of some game, say baseball, the promotion of an ideology? This book is simply an effort to set forth the express or tacit rules of science as a game. If a sociologist does not like the rules, easy, don't play.

Finally, we hope that readers will recognize that space limitations precluded treatment of many important subjects in sociology. One of the glaring omissions is "expectation states theory," centering on the publications of Professor Joseph Berger and his numerous collaborators. Even more glaring, we create the grossly misleading impression that sociology is practiced only in the United States. With these caveats in mind, let us begin.

Sheldon Ekland-Olson, Rapoport Centennial Professor, University of Texas at Austin
Jack P. Gibbs, Centennial Professor Emeritus, Vanderbilt University

What does a scientific sociology look like? What role does uncertainty play in the search for truth? Do the social sciences have something unique to teach us about science in general? More than a breath of fresh air, this book is a breath of life for a field that's been suffocating increasingly in the grip of special interests and identity politics. Ekland-Olson and Gibbs are at the top of their "game" (their term) here: in clear, precise, and cheerful manner they demonstrate what is called for in that highest and rarest act of predictive theory-building.

HC Hsu, PhD and Fellow, Europäische Universität für Interdisziplinäre Studien

Olson's and Gibb's *Science and Sociology* promises to reinvigorate attention to classical issues in sociological theory and methods, including basic concepts. They begin with a discussion of the centrality of "assertions" to science and end with advocacy of "predictive power" as the criterion for defending theories. Olson and Gibbs cover a multitude of definitions of science and convincingly defend their preferences. In addition, we are coached in the fundamentals of formal theory construction, including the works of a wide range of sociological scholars in the process. The work should generate lively discussions of sociological epistemology.

Gary F. Jensen, Professor Emeritus of Sociology, Vanderbilt University

In an era when public trust of science has eroded, Ekland-Olson and Gibbs provide a much-needed renewal of the scientific foundations of sociology. They relish in the importance of uncertainty as the key ingredient in the scientific enterprise of sociology, and persuasively place theory testing at the heart of the advancement of scientific sociology. Their goal is to elevate sociology's scientific status, and this book provides an essential roadmap for doing so.

David Kirk, Associate Professor of Sociology, Professorial Fellow of Nuffield College, University of Oxford

1

ASSERTIONS

The Building Blocks of Science

We begin with the subject of "assertions," the basic "stuff" of science. Assertions are not just the building blocks of science; they also promote its progress. That contention may appear far from radical, but it points to a crucial distinction.

Assertions are not facts. Rather, as used throughout this volume, an assertion is a claim, infused with varying degrees of certainty, about an event or thing (past, present, or future), perhaps a claim about a category or class of such events or things. The claim may be about the existence or nonexistence of a characteristic. It may also be "relational," which is to say a claim about the spatial or temporal linkage between or among characteristics, be they quantitative or qualitative.

The identification of an assertion as science's subject matter may appear radical to some, as we frequently talk about science as dealing with "facts." The conventional picture is that of a scientist searching for a pattern or some uniformity, one that he/she wants to report to peers and perhaps to a broader audience. Yes, searching for patterns and subsequently reporting them are fundamental scientific activities, though it misleads to describe the discovered patterns as "facts." They are assertions, accompanied by more or less empirical certainty, a quality that sets them apart from recognition of ignorance. This distinction is so far from trivial that it warrants repeating another way. Scientists communicate their beliefs, generalizations, and theories in the form of assertions *not* facts. Consequently, science is both adversarial and progressive.

The Only Certainty is Uncertainty

What separates assertions and facts (as commonly understood) is uncertainty. However, those writers who make "facts" the central notion in science (David Wootton [2015] being the latest) are not prone to offer a definition of "a fact," if any at all, other than a version of "something—a thing or an event—known or believed with certainty." The immediate question: how many must know or believe in that "something" for it to be a fact? Surely not everyone, but any lesser number or proportion would be screamingly arbitrary. Furthermore, although such a definition makes "a fact" private knowledge, we (the authors of this book) defy anyone to formulate a more defensible alternative. Should Wootton take exception, in writing fifty-nine pages on "facts" (2015: 251–309) he did not set forth an explicit definition of a fact (that is, his own). Moreover, he did not use the word "events" as well as "things" in his seemingly endless description of *attributes* of facts (e.g., by definition true, trump card in an intellectual game, not just true or false, based in experience). In writing on those attributes Wootton did not recognize that a series of historical observations, no matter how long or verifiable, does not constitute a definition.

Now consider an illustrative implication. It is likely that you—the reader—knows with absolute certainty that you were never in Holder, Texas; but it is also likely that no one else knows of your ambulatory history, definitely not with absolute certainty. Accordingly, in that "something" we have a fact, and like uncountable others it has no scientific relevance whatever if only because it is *private*. By contrast, assertions are public, and science is a public institution.

It is widely held that the surest path to certain, some would say "factual," knowledge of how the world works is through carefully controlled experiments. The immediate and foremost problem is that experiments are not always possible, nor ethically advisable; and that is a major limitation for sociology, along with other social sciences. For that matter, whatever the science, regardless of how careful, some slack is always possible when interpreting an experiment's outcome. As David Lindley (2007: 2) has argued:

> [in the] back-and-forth between experiment and theory, it's uncertainty that tells the scientist how to proceed ... When scientists have resolved one level of disagreement, they move to the next. Uncertainty, discrepancy, and inconsistency are the stock-in-trade of any lively scientific discipline.

Relatedly, Darwin argued that, to be constructive for scientific purposes, observations must be antagonistic or partisan, for or against something. Even so, when observations are made to refute or support some generalization, both the observations and the generalization are assertions. Such is the nature of the scientific enterprise, no matter what scientists or their critics say.

Those who hold that science is simply embedded in "neutral observations" of events or things may take exception to Darwin's claim. We might, for example, be simply interested in the number of leopards ostensibly present in a designated African locale as of some particular period. Likewise, we might simply be curious about how many people voted in the last U.S. presidential election. Whether adversarial or neutral, the methods employed, the name of the observer, the purpose of the study, or how findings are disseminated, the observations reported by scientists and their generalizations are assertions accompanied by greater or lesser certainty; nothing more, nothing less.

The Importance of Uncertainty

Variation in the "uncertainty level" of assertions is endemic to both the substance and progress of science. For those familiar with the history of science this assertion will not be startling; indeed, it may sound all too familiar. For a few it will bring to mind the young Werner Heisenberg working in the 1920s on the implications of the quantum revolution ignited by his senior colleagues Niels Bohr, Albert Einstein, Paul Ehrenfest, and Max Planck. Noting unanswered questions and irregularities in findings, Heisenberg proposed what came to be known as the uncertainty principle, profound in implications and simple to state. When measurements are of the approximate same wavelength as that being measured, uncertainty results. Stated more succinctly, the kind of observation instrument affects the measurement taken. So stated, the uncertainty principle holds whether we are focused on the subatomic phenomena of quantum mechanics, the field notes of an ethnographer in Brazil, or responses being gathered by a survey researcher in Minnesota for the National Longitudinal Surveys.

Lest we despair, finding ourselves adrift in a sea of uncertain assertions by scientists, note especially that even in the face of uncertainty, predictions based on *assertions* can prove quite useful. Indeed, scientists are fond of referring to "five-nine certainty—99.999%." With that certainty level, much can be accomplished. As a case in point, the assertions of quantum mechanics have given us lasers, semiconductors, designs for cancer drugs, and quantum computing. At far lesser levels of certainty, theories about crime or criminality are invoked to support this or that policy, and the implications of increasing life expectancy are considered in contemplating future demands on health care.

Assertions, Definitions, Applications, and Tests

While the fact/assertion distinction is our beginning point, an additional distinction warrants recognition. Definitions are also of critical importance to science, but they are not assertions. For example, the statement "Bachelors are men who have never been married." is true by the very meaning of the constituent words. If one assumes that the man never married, the statement is simply a definition. But to say "He is a bachelor because he grew up with a domineering mother" is to state a debatable assertion. In that connection, philosophers are fond of questioning the distinction between analytic and synthetic statements, as though it somehow poses an insoluble problem. To the contrary, it is difficult to see how anyone can deny that the distinction is at least fairly clear. A synthetic statement is one whose truth/falsity is *not* determined by the meaning of the constituent words/or and syntax. In contrast, an analytic statement is a statement that is true or false by the very meaning of the embedded words and syntax.

Despite the foregoing, it should be noted there is some long-standing disagreement regarding what Quine (1951, 1960) has called the two dogmas of empiricism. The position taken here is that if, as Quine evidently contended, the analytic/synthetic distinction is a delusion, then several trainloads of scientists have wasted much of their career attempting to test or conduct some other kind of research on what they thought of as synthetic statements. Alternatively, if Quine simply recognized that the distinction is difficult and/or debatable in many instances, his argument is true but trite. This debate need not long detain us here; but for reasons set forth in subsequent chapters, the analytic/synthetic distinction becomes especially important when testing generalizations or theories.

Consistent with the foregoing, the tendency of science, indeed its goal, is to move forward in a systematic fashion, with consensual definitions of words and terms that make scientific generalizations qualify as synthetic statements and possibly testable. Described another way, generalizations are assertions of an empirical association or condition, and they apply to more than one thing or event, perhaps an infinite category (or "class"). For example, consider the assertion: "No hunting/gathering population is socially stratified." To assess that assertion, we need definitions of "hunting/gathering," of "populations," and of "socially stratified." Now comes the rub. We need to apply these terms across locales and do so consistently with the agreed-upon definitions. The application problems are not always easily resolved, and they can be conducive to heated debate, especially when asserting a negative, as above. Such debates are all too rare in sociology, though that rarity hardly indicates that assertions do not flourish in the field. To the contrary, only their seeming indifference to testability prevents sociologists from recognizing that they are a fount of assertions.

More on Assertions

In the context of scientific activities an assertion is a scientific statement in the way of a claim, one set forth with the express or tacit understanding that it will be judged—accepted or rejected—in light of data gathered to assess its predictive power relative to that of contenders. In making such assessments, scientists need to *connect* definitions with data collected for assessment purposes.

Epistemic Statements

The connections in question are themselves assertions in the form of epistemic statements, meaning statements that connect words, terms, or definitions with ostensibly observable events or things (past or present, particular in any case), possibly numbers of some kind. An illustrative instance: "Between 1890 and 1990 the amount of CO_2 in the atmosphere rose … from 285 to 345 parts per million by volume" (National Geographic, *Concise History of Science and Invention*, 2010: 331). The sentence actually implies three epistemic statements (ESs), all pertaining to "referents" (here numbers) as characteristics of "amount of CO_2 in the [Earth's] atmosphere" as of 1890, 1990, and amount of change 1890–1990. All three of the implied ESs would be more complete by stipulating how the referents were created (e.g., the kinds of instruments or devices employed), which is to say that an ES is not truly complete until extended to an identification or description of the source/origin of the referent or referents. However, no matter how elaborate or plausible, ESs are assertions and, hence, disputable; and the contentions may not be limited to the obvious possibility of mistakes in calculating numbers. For that matter, opponents of the "global warming argument" could go so far as to allege that the CO_2 numbers were knowingly inflated. Despite such possibilities, the referents could be used in testing some theory about climate change.

There is nothing esoteric about the notion of ESs, and they are definitely not peculiar to science. So we hear someone say: "Fred told me that Jimmy is out of town." We will have heard an epistemic statement, and like many others it is a "dual assertion," both disputable. Fred may have said no such thing and, even if he did, Jimmy may still be in town. So what we heard is simply an assertion, as all ESs are.

Should it be objected that the Fred/Jimmy illustration blurs the distinction between an assertion and the "meaning of a statement," the two cannot be separated neatly, if at all. The problem is the suggestion that all assertions should be questioned. Perhaps so; nevertheless, it is obvious that human beings, surely including scientists, routinely accept all manner of assertions without doubting them for a moment.

The ultimate problem: there is no final, indisputable way whereby ESs come to be accepted in science. The kind of connection in question is bound to entail debatable assumptions and a ubiquitous source of possible error, especially in the case of purported measurements or estimates (e.g., the age of Earth). So what is regarded as "scientific knowledge" is really a study in assertions, and they include a vast array of ESs. The point is that one version or another of an epistemic statement may come to be accepted in some scientific field (e.g., geology) to the point that it can be thought of as an "application rule," meaning a conventional way of using some word, term, or definition to identify or designate some observation or datum. The reference to an application rule points to a distinction between ESs accepted by scientists and mundane ESs. Think of someone repeating what Fred said about Jimmy. No one is likely to think of that repetition as in any sense suggesting a rule.

The Larger Picture

The foregoing contentions are clarified in subsequent treatments of definitions, generalizations, and theories (Chapters 2–7). The more immediate need is a depiction of science in general.

The Elements of Science

To repeat, in "doing" science assertions are made, definitions developed, measurement strategies devised, and tests conducted; but the links always remain somewhat uncertain. The four elements of the progressive scientific enterprise are depicted in Figure 1.1, though in full recognition that the representation is a gross oversimplification.

A definition that accompanies an assertion is empirically applicable to the extent that independent investigators agree when classifying or describing (including measurements) characteristics of the phenomenon defined, the phenomenon being the *definiendum* (e.g., Mars, planets, middle class, a city). The exact degree of empirical applicability is always somewhat uncertain, and it is often entangled with the reliability and/or validity of the measures used in the case of quantitative characteristics. Library stacks have been filled and a mountain of Internet postings made treating reliability/validity pertaining to empirical applicability, though that term is rarely used in assessing some proposed definition or research procedure.

For now, it is enough to state that the testability of any given assertion cannot be known with absolute certainty (actually a matter of consensus), a problem that haunts all sciences. But some good news: the problem has proven less an impediment than worried philosophers suggest. Regardless of the embedded uncertainty, it is through convention, development of instruments, and debate that scientists are able to arrive at some measure of effective agreement on the meaning of particular empirical assertions and the procedures for testing them.

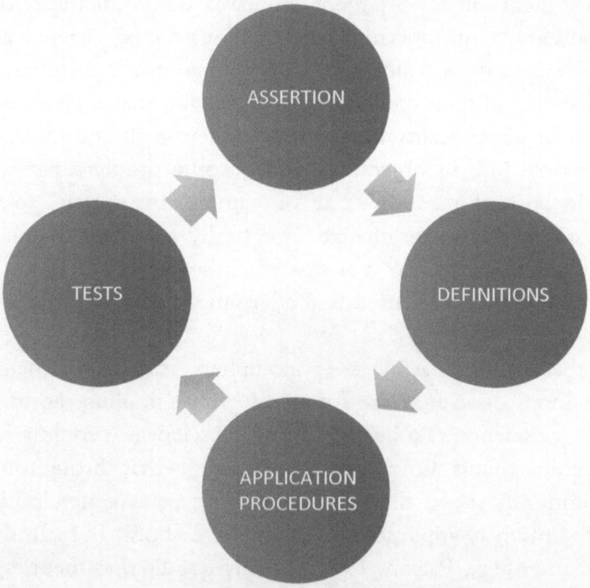

FIGURE 1.1 The Progressive Cycle of Science

When at least an approximation of such agreement is realized, we have what Thomas Kuhn (1996 [1962]) called "Normal Science." Such science prevails until disagreements emerge in the form of new assertions and/or different interpretations, with or without shifts in test procedures. Such changes prompted Kuhn to write about "paradigm shifts and scientific revolutions." Setting aside Kuhn's vague use of the word "paradigm" (Masterman 1970), it can be safely said that the progress of science proceeds along a jagged and frequently contentious path, replete with novel and important assertions, frequently infused with uncertainty and controversy. The result is a journey guided by commonly accepted assertions and procedures, though dramatic and unexpected shifts in direction are always possible.

Whatever the "language problems" of science, empirical applicability being the major one, they have not precluded, as some philosophers are prone to imply, achievements in science. These achievements contradict any claim that science cannot overcome some alleged insurmountable problem. Instead, the inherent uncertainty of science underscores the assertion that the essence of the scientific enterprise is constant dispute, controversy, and frequently successive but temporary resolutions. To be sure, as science proceeds some of its achievements may prove ephemeral, though it is foolhardy to claim that the scientific achievements have no more substance than folklore or other social constructions external to science. That point would not be worth making if it were not for the contrary contentions of postmodernists and deconstructionists, which all too many sociologists have swallowed hook, line, and sinker.

Notable Naysayers

Building on the work of Friedrich Nietzsche, Paul Feyerabend (1975) developed what has come to be known as epistemological anarchy, arguing at one point that "anything goes" when it comes to scientific methods. Feyerabend is not alone (Kuntz 2012). According to these naysayers, the idea that science attempts to enlarge the boundaries of what we *objectively* know is a pipe dream.

Working in the context of an emergent emphasis on multiculturalism and a resulting interest in moral relativism, postmodernists have "deconstructed" science and asserted there is no universal truth. That assertion is not totally alien to admitted uncertainties in science, but those critics of science extend their contentions to the claim that each social or political group has the right to assert whatever reality best promotes their goals and interests (Otto 2011). By that route, science would lose its objectivity and become simply a pawn in political battles over policy. It would be nothing more than one among many ways to frame the truth, as witness current debates over climate change, genetically modified organisms, and the social implications of same-sex marriage. So it is that the inherent uncertainties of science can be, have been, and will continue to be used to undermine science's authority, perhaps even its legitimacy.

The question for the scientific naysayers—philosophers, academic humanists, or otherwise—is this: if the uncertainties of science are somehow truly crippling, again what of the indisputable achievements of science? To be sure the social sciences, certainly sociology, have few if any indisputable achievements. One very likely reason is that throughout the field's history sociologists have commonly stated their assertions, perhaps as generalizations or theories, as though indifferent to empirical applicability, testability, or both. Instead, they have promoted broad conceptual schemes (e.g., Parsons 1951), usually disguised as theories, or more narrowly focused exercises in "explaining by naming" (e.g., Goffman 1963). If those publication dates are far too distant, we invite readers to scour more recent sociological literature until exhausted, trying to find reports of research on the empirical applicability of any major sociological notion (with the possible exception of the attention given to clarifying "social class"). However, there is no mystery. There are many roads to tenure in a sociology department, but research on the empirical applicability of some term is not one of them. More on the subject of empirical applicability comes in subsequent chapters, much more.

For the moment it suffices to recognize that critics are likely to dismiss the foregoing references to "scientific achievements" as painting an idyllic picture. To which we respond: how could an emphasis on inherent uncertainty possibly qualify as "idyllic"?

Generalizations and Theories

We come now to a final question before diving into more detailed treatments: what is the relation between scientific generalizations and theories? We might say that a theory is a collection of generalizations, but that simplification misses a conventional and important distinction drawn between two major substantive divisions of a theory—its premises (axioms, postulates, and/or propositions) and its conclusions (theorems). When looking for a theory we seek a set of assertions wherein some testable generalizations are systematically derived from those that cannot be tested in anything like a direct, proximate sense.

An example might help. Seeking to avoid charges of misconstruing or otherwise mangling the ideas of others and knowingly opening ourselves to direct criticism, we draw upon the work of one of the authors. Some fifty years ago, a theory about the suicide rate was proposed (Gibbs and Martin 1964), based on the idea that life is filled with tensions, disruptions, and even grief, all stemming in large part from a lack of stability and durability in social relationships. From this idea, five *untestable* postulates were formulated, as follow:

1. The suicide rate of a population varies inversely with the stability and durability of social relationships within that population.
2. The stability and durability of social relationships within a population vary directly with the extent to which individuals in that population conform to the patterned and socially sanctioned demands and expectations placed upon them by others.
3. The extent to which individuals in a population conform to patterned and socially sanctioned demands and expectations placed upon them by others varies inversely with the extent to which individuals in that population are confronted with role conflicts.
4. The extent to which individuals in a population are confronted with role conflicts varies directly with the extent to which individuals occupy incompatible statuses in that population.
5. The extent to which individuals occupy incompatible statuses in a population varies inversely with the degree of status integration in that population.

The postulates imply this testable theorem: **The suicide rate of a population varies inversely with the degree of status integration in that population**. Status integration is measurable because it is simply the proportion of some population (e.g., all women aged 40–44 residing in a specific city) who occupy a particular achieved status (any particular one). Thus, if .102 of a population are "divorced," that is the degree of integration for that particular marital status. It follows that the degree of *marital integration* for a population is the extent that its members are concentrated in *any one* of the marital statuses, and that extent is the sum of the squares of the proportions (one such proportion for each marital status). What has been said of marital statuses applies to all other kinds of *achieved* statuses (e.g., occupations).

Note especially that the theory does *not* assert a direct connection between status integration (SI) and the suicide rate (SR). Rather, it asserts that the SI/SR negative association holds only through the postulated direct relation between SI and the stability/durability of social relations. That assertion is a translation of a metaphorical language to an empirically applicable language, a translation that points to a common crippling defect of sociological theorizing. Describing it all too briefly, sociologists appear unable or unwilling to translate metaphorical statements into testable statements.

Tests of the theorem have been reported in numerous publications pertaining to several kinds of variation in the SR (see Gibbs 2001 for references to fifteen publications pertaining to the theory, test outcomes in particular). It is especially important to recognize that the theory generates *testable* predictions about various kinds of variation in the SR (e.g., among age groups, among marital statuses within an age group, among occupations) that cannot be deduced systematically from any contending theory (see references in Gibbs [2001] to the theories of Henry/Short and Durkheim). Such a contrast pertains to the "range" of theories, a dimension of predictive power examined more fully in Chapter 4.

The theorem is testable because both "status integration" and "suicide rate" can be defined by a formula that applies to published and, hence, readily available data. As for the deduction of the theorem, think in terms of three quantitative variables—*X*, *Y*, and *Z*. Now assume the following two relational assertions: the greater the *X*, the greater the *Y*; and the greater the *Y*, the less the *Z*. Even if only *X* and *Z* are empirically applicable, in light of what is known as the "sign rule" (explicated in Chapters 6 and 7), the two assertions imply this testable generalization: the greater the *X*, the less the *Z*. Unfortunately, the vast majority of sociological theories are not stated in a form that makes such systematic deduction feasible, and the consequence is more than a century of untestable theories. That outcome reflects the seeming determination of sociologists to assess theories in terms that have no bearing on testability (e.g., intellectual satisfaction, plausibility, elegance). Until this shortcoming is rectified, progress will be slow, if forthcoming at all.

Other Implications

Returning to the SI theory, it implies more than the one relation. Specifically, it implies that among territorial units (cities, metropolitan areas, states, or provinces) there is a positive association between the SR and each of the following population percentages: the unemployed, the widowed, the divorced, and new residents (i.e., migrants of one kind or another).[1] Why those implications? Because the degree of SI is *less* for those statuses. Moreover, each status can be construed as involving one or more disrupted social relations.[2] Such involvement surely has some connection with the notion of "strong and durable social relations." As in the case of other metaphors that pervade sociology's vocabulary, it is difficult to imagine any defensible and feasible measurement procedure, the principal reason why Postulate 1 (like any other assertion so labeled) is considered untestable in any direct sense.

Finally, it is recognized that, whatever the territorial units, the associations in question are so-called "ecological correlations" (more accurately, aggregate correlations). Consequently, it is desirable to speak to the usual bromide in the form of this assertion: ecological correlations may not hold at the individual level. Because of space limitations, we—the authors—must be content to point out that a mountain of research findings (numerous countries and various periods) clearly indicate that the unemployed, the widowed, and the divorced do have substantially greater SRs than do counterparts. The only doubt pertains to "new residents" (migrants of some kind), but we do not know whether official SRs have ever been computed for such a population.

Varying Forms of Assertions

From all of the above we might conclude that "evidence" is the most basic notion in science. While the importance of objectively gathered evidence cannot be exaggerated, it always comes in the form of assertions. Nonetheless, while assertions do not reduce science to a debating club, something akin to debate pervades science and clearly fuels progress. Indeed, the ancient Greeks' appreciation of something akin to democracy in particular population divisions (surely not for slaves) and certainly their development of rhetoric may well have been their greatest contribution to science.

Expressing assertions is only a *necessary* activity for science. Were it both necessary and sufficient, then social commentary, journalism, practicing law, and doing history would be sciences. Even among sociologists and their close academic cousins—anthropologists and political scientists—assertions about how the social/political world works come in many forms. Some aim to shed light on specific social situations and how the social construction of reality unfolds (e.g., Berger and Luckman 1966). In such situational assessments there is commonly little attention given to the formulation of generalizations beyond something like this: Our beliefs, actions, and sense of identity are shaped by the situations we encounter. Unfortunately, that generalization is far from testable, insufficient empirical applicability being the immediate problem.[3]

Some Strategic Illustrations

In the early 1970s anthropologist James Spradley studied the life of urban nomads, those more familiarly known as the skid row homeless, concluding that "The urban nomad culture is characterized by mobility, alienation, and a unique set of survival strategies … Whatever the initial impetus, once a man moves to the edge of this world he will be thrust to its center by repeated incarceration." (1970: 253) Spradley surely made assertions about life among "urban nomads," but his primary aim was simply to promote better understanding of this world of "strangers and friends," not readily testable generalizations about infinite categories or classes.

More recently, after observing and gathering data on a shantytown in Argentina, Javier Auyero (2009: 159) reached this conclusion:

> [We] have presented many individual testimonies that speak about suffering. The experiences of affliction, however, are not simply individual. They are social because, although located in individual bodies and expressed in individual voices, they are actively created by the position that … residents as a group occupy as materially and symbolically deprived persons …

As compelling as Auyero's description may be, it is by no means clear what it implies about *other* populations, especially in the way of testable generalizations.

As another illustration of the "we become what we are in the situations we encounter" perspective, consider the research of Philip Zimbardo (1972) and his colleagues (Haney, Banks, and Zimbardo 1973). They conducted an experiment at Stanford University on the "pathology of imprisonment," in which some students played the role of prisoners and others the role of punitive guards. The experiment was eventually terminated because of ethical concerns, stemming from evidence that the participants came to perceive themselves as doing more than playing a role. Three decades later in the *Lucifer Effect* Zimbardo (2007: 321) summarized the experiment in the following statement:

> Our sense of identity is in large measure conferred on us by others in the ways they treat or mistreat us, recognize or ignore us, praise us or punish us. Some people make us timid and shy; others elicit our sex appeal and dominance. In some groups we are made leaders, while in others we are reduced to being followers. We come to live up to or down to the expectations others have of us … Those subjective beliefs create new realities for us. We often become who other people think we are, in their eyes and in our behavior.

The list goes on and on, and the social science literature is replete with such assertions drawn from situational assessments. Some are generalizations with clear predictive implications, but others are simply descriptive or interpretive of particular situations (e.g., Erickson 1976, Anderson 1990, Bestor 2004, Desmond 2007, A. Goffman 2014).

Assertions Made to Promote an Ideology

Ethnographies and field experiments are not the only sources of loosely formulated assertions in the social science literature. They come also from *interpretations* of history and particular perspectives (e.g., Freudian, Marxist). One outstanding instance is Professor Joe Feagin's presidential address to the American Sociological Association (ASA) in 2000. Feagin built his call for increased attention to social justice on an assertion by Karl Marx—that modern capitalism creates bad economic times, encompassing both social injustice and inequality.

The newly installed president of the ASA gave several examples of the "troubling conditions currently being created or aggravated by modern capitalism." Among these are "many of the world's people still live in misery," working families are "exploited and marginalized," and capitalism "generates profits at the huge cost of increasing environmental degradation" (Feagin 2001). The problem with such assertions is they are difficult to test in any concrete way, as distinct from claiming that the assertions are obviously true and dismissing naysayers as naïve.

Before accepting Feagin's assertions, numerous questions await answers. Are the conditions unique to capitalism? If so, what aspects of capitalism? How do we know that rates of personal misery, the exploitation and marginalization of working families, and degradation of the environment do not flow, perhaps at even higher rates, from any number of alternative systems for organizing markets?

Lest this critique of Feagin's address be construed as a defense of capitalism, it is nothing more than a call for clarity. Without greater clarity, there is simply no systematic way to test Feagin's generalizations and thereby support or refute them.

Other Forms of Assertions

Many, more precisely formulated assertions come from survey researchers and demographers. Building from a large body of data, Bengtson and Roberts (1991) advanced a theory about "intergenerational solidarity in aging families." It comprises, inter alia, careful definitions of such terms as affectual solidarity, normative solidarity, and structural solidarity. The authors (1991: 857) rightly emphasized that the theory emerged from "successive iterations between conceptual statement and empirical verification."

Likewise, among demographers, in what has become a classic article, Kingsley Davis identified several factors in the "world demographic transition." His description of that transition includes both testable generalizations and policy statements, a demonstration that scientific statements and value judgments may appear in the same publication. Consider the following passage (Davis 1945: 11):

> As the cases of Russia (in retrospect) and India (in prospect) illustrate, it is virtually impossible to make the transition from an agricultural to an industrial regime without dislocating and disorganizing great sections of the population … There is ample room and indeed a great necessity for the development of a scientifically grounded population policy throughout the world.

From such publications and earlier work by Warren Thompson (1929), a five-stage "demographic transition model" emerged. Over subsequent years a mountain of research was designed to test, revise, and refine the model's predictive implications (e.g., Caldwell 2010).

A Distinctive Line of Theorizing

Finally, we note the sweeping generalizations of what adherents refer to as "pure sociology," grounded firmly in the work of Donald Black, who writes in *Moral Time* (2011: 4–5): "The fundamental cause of conflict is the movement of social time." More specifically, "The greater and faster the movement of social time, the greater is the likelihood of conflict and the greater the conflict is likely to be." *Moral Time* extends Black's earlier work, *The Behavior of Law*, which stimulated considerable research, including arguments over measurement. The outcome has been both confirmation and contradiction of Black's generalizations (e.g., Gottfredson and Hindelang 1979a; Hembroff 1987; Lessan and Sheley 1992; Bernard 2002).

Black recognized that his theory is neither perfect nor final, with a special and glaring need, if testability is the goal, for careful empirical specifications of "social time" and the "shape of social space" (among other terms). Whatever the measurement strategies devised and eventual refinements made, the theory's claims are sweeping:

> Social time causes conflict in all human relationships—between family members, lovers, friends, acquaintances, strangers, organizations, tribes, and nations.
>
> *(Black 2011: 6)*

A Quite Different Kind of Survey

What we find in the sociological literature is a variety of kinds of assertions. Some are isolated with few if any obvious predictive implications, some are predictive hypotheses, some compare contradictory predictions, and some are ostensibly theories that comprise very broad generalizations.

Out of curiosity and with no claim to representing the variety of sociological publications, we drew a random sample of 100 articles from the *American Sociological Review* (ASR) published during the decade 2004–2014. We wanted some sense of the kind of assertions made in the ASA's cornerstone journal.

About a third of the articles reported isolated assertions, with little attention, if any, to predictive implications. Just under 50 percent of the articles reported tests of isolated hypotheses, with some discussion of what the findings pointed to in the way of further research. No more than one-fifth of the articles reported tests bearing on the relative predictive power of *contending* generalizations. Only two articles over this decade presented a distinct set of generalizations in the form of a multivariate theory, generalizations presented as variables in a regression equation.

This all too brief survey of ASR articles suggests a very mixed bag. There is some sense of assertions as ingredients of science but not as somehow essential. Similarly, several of the publications clearly suggest that sociology is an empirical enterprise, but there is surely no emphasis on the need to formulate testable theories. Indeed, there is scarcely an emphasis on the need for theories, whatever their testability.

Some Bottom Lines

The position taken here: although assertions void of predictive implications may shed light on important topics, an assertion is fully scientific if and only if made in the understanding that it will be assessed, directly or indirectly, in terms of its predictive power (dimensions of which are described in Chapter 4). Correlatively, scientists are prone to react to an assertion by contemplating, proposing, or actually undertaking an assessment of it in terms of its testability and predictive accuracy. Stated another way, social constructivists and postmodernists notwithstanding, scientists take the idea of gathering theoretically relevant but objective data seriously.

Sociology's scientific status is debatable, though not because practitioners rarely make assertions. Many sociologists stand proud at center stage asserting with gusto but rarely displaying genuine concern with testability. They react to an assertion by analyzing, condemning, or praising, which is to say anything other than testing.

But do not construe our reservations about the current scientific standing of sociology as manifesting the bitter end of long careers as sociologists. At the risk of being politically incorrect, we are like the Cuban who said "Baseball's been good to me." More important, we write in step with the admonition "Keep hope alive."

Notes

1 The associations may be much more substantial when the suicide rates are three-year average annual. Moreover, the association between any one of the variables (e.g., unemployment rate) and the suicide rate is expected to be substantially greater when the other variables are statistically controlled in the form of a partial correlation coefficient or a regression coefficient.

2 Needless to say, it is certainly questionable to assume that all kinds of migration (or "residential mobility") involve disruptions of social relations, let alone equally as regards extent and intensity. So there are more doubts about the association between suicide rates and residential mobility than about the other associations (i.e., with the percent unemployed, divorced, or widowed). There are also doubts about the "suicide/never married" association, but in that case the doubt stems from recognition that "never married" does not necessarily involve disrupted social relations.

3 As opposed to the term "empirical applicability" itself, scientists in all fields, save some of the social sciences and sociology in particular, are fully aware of the *notion* of empirical applicability and its importance. Savor Simon Barnes' statement (2014: 36) "Classification is not definitive: it is about agreement, and science is not conducive to agreement." Barnes could have added that in the advanced sciences definitions are vulnerable to rejection (a manifestation of disagreement) on the grounds of demonstrable empirical *inapplicability*. Sociologists do disagree about definitions, but the reason commonly has little to do with empirical applicability.

2

ON PREDICTIVE IMPLICATIONS

To say it again, assertions are the building blocks of science. They imply predictions about how the world works, but they are infused with inherent uncertainty. This uncertainty emerges from the ideas pursued, the measurement strategies employed, and the stochastic nature of the subject matter investigated. So viewed, science is an attempt to produce a body of knowledge, though inevitably uncertain when judged by the degree of testability and predictive accuracy. Seems clear enough. There is, however, some remaining fog. In this chapter we take a brief look at the distinction between empirical applicability and logic, especially conceptual clarity.

Logic and Empirical Applicability

It was Albert Einstein, in an address to the Prussian Academy of Sciences (1921), who said: "As far as the laws of mathematics refer to reality, they are not certain; and as far as they are certain they do not refer to reality." It does not distort to say that Einstein was talking about assertions and the inevitability of uncertainty in science, even though Einstein himself occasionally appeared very certain about his own assertions.

So how do scientists cope with uncertainty about assertions? They test them in that they (1) expressly or tacitly examine a given assertion with a view to identifying its predictive implications and (2) engage in some activity to ascertain the extent that some particular implication is realized. The crucial point is that an assertion cannot be tested unless at least one of its predictive implications can be identified with confidence. Two points of confusion frequently emerge, one centered on the logic employed and the second on the empirical applicability of the terms in the assertion.

Logic and Predictive Implications

Obviously, a test of an assertion based on future data or future observations is a literal impossibility, as all data and observations pertain to what can or has been observed in the present or past. Accordingly, be it a component of scientific theory or an isolated generalization, an assertion has no *necessary* connection with prophecies, foretelling, or extrapolation in the conventional meaning of those words. Rather, an assertion may pertain to the future in the sense that it implies a prediction pertaining to a "conditional future" as opposed to a "literal" future. Stating the conditional abstractly: *if* such-and-such happens, such-and-such will follow. Stated still another way, if properly trained and informed investigators engage in such-and-such an activity they will report having experienced such-and-such. The first "such-and-such" is conventionally described as "observational behavior," which may or may not involve the use of devices (instruments or machines); and the second is conventionally known as "perceptual behavior." However, both are components of a predictive implication.

Consistent with the foregoing, an assertion's predictive implications are revealed, at least partially, by an answer to this question: what empirical outcome can be deduced from the assertion by some designated logic? Whether inductive or deductive, logic is nothing more or less than a set of rules—perhaps only one—for relating statements, propositions in particular. The rules of the classical syllogism and those of mathematics are widely accepted instances. However, mathematical models and logic schemata per se have to do with the relation between and among propositions, *not* with tests of the empirical implications of assertions (all "empirical" or "contingent"). This seemingly simple point sometimes gets lost.

Consider this straightforward assertion: Abraham Lincoln was born in Kentucky. Although simple and straightforward, the assertion cannot be assessed defensibly on the basis of logic alone, and the complexities, even in this apparently elementary assertion, defy exaggeration. In particular, it might be assumed that the name "Abraham Lincoln" will surface in the records of the 1810 census for Kentucky, but no conventional logic supports that assumption (i.e., the assertion does not imply the assumption or, stated otherwise, the assumption cannot be deduced from the assertion). Moreover, it is possible that the person recorded in the Kentucky census of 1810, assuming names were recorded, as "Abraham Lincoln" did not become President of the United States (the possibility brings to mind an amusing assertion: Homer was not the author of the *Iliad*; rather, it was some other fellow named Homer). Even assuming that only one "Abraham Lincoln" was born in 1809, regardless of any reported date, Lincoln's actual birth could have been after the 1810 census. For that matter, he could have somehow simply been overlooked in that census (another amusing expression comes to mind, one attributed to Mae West: Better to be looked over than overlooked).

The point is as simple as it is important. The distinction between *explicit* assertions and *implied* assertions—observations, reports, or other data—should always be recognized. Nevertheless, despite that recognition, scientific assertions must be explicit and directly or indirectly testable, and in either case there must be agreement that a reported test of an assertion questions (if not totally refutes) or supports some predictive implication of the assertion. Thus, returning to Feagin's presidential address to the American Sociological Association referred to in Chapter 1, there is no denying that some of his statements are assertions, but they are not testable because their predictive implications are obscure.

Pursuing difficulties associated with the notion of predictive implications a bit further, consider the *empirical* implications of a "logic" so preposterous that no one would take it seriously: whatever the property or attribute in question, if any member of any class of events or things has or had that property or attribute, then all members have or had it. Accepting the assertion that Lincoln was assassinated, that astonishing logic of induction implies this prediction: an examination of biographies of each U.S. President will find evidence that all were assassinated.

It will not do to dismiss the foregoing "logic of induction" as uniquely preposterous; the problem is much more general. Try to think of any logic that would make it possible to deduce from any premises a correct prediction as to *anything* about the assassination of U.S. Presidents. Extending the challenge, try to identify any logic that generates putatively correct answers to genuine empirical questions. To repeat the point, perhaps *ad nauseam*: Logic has nothing whatever to do with the predictive accuracy of any particular assertion; rather, it consists of rules as to, *inter alia*, what any particular proposition implies in the way of other propositions. In that connection, were it not for the vagaries inherent in natural languages (e.g., Russian, English, French, Chinese), it is conceivable that logicians would never have invented the classical syllogism; and scientists might never have come to rely on mathematics as much as they do.

Other Illustrative Arguments

Some arguments appear reasonable to the point of being totally acceptable, even though not based on a fully articulated logic. Indeed, humans habitually argue in terms of an "implicit logic," and scientists commonly report tests of assertions without fully disclosing the underlying logic, whatever it may have been. Perhaps it was only a hunch, which is to say "nonlogical." Indeed, the practice is virtually an art form in sociology. Nevertheless, identification of an assertion's predictive implications on the basis of a less than fully explicit logic (if one at all) is questionable. Brief illustrations may be helpful.

A Very Clever Man's Argument

Bertrand Russell once described "experience" as a vastly overrated notion. He said something like "I have no experience of ever having been born, but I would not doubt it for a minute."

Russell could be interpreted as in effect having asserted "I was born." But he actually made an enthymematic argument, one with at least two missing premises. First, all human beings were born. Second, I am a human being.

A Quite Different Argument

Talcott Parsons claimed (Parsons and Shils 1951: 49) to have carried "deductive procedures further than is common in the social sciences." Yet it is widely agreed that his action theory is amorphous if only because it is difficult to detect any *deductive procedures*. As stated by the philosopher Max Black (1961: 271): "There is very little strict deduction in Parsons' exposition." And bear in mind that Parsons was once the leading theorist in sociology.

More recent distortions of "logic" by sociological theorists are not readily found, largely because the arena of sociological theories, testable or not, has become a desert. Perhaps it is just as well until sociologists adopt some mode of formal theory construction, a subject we will pursue in Chapter 6.

A Second Clever Man

A final illustration of problems with the notion of "implication" comes from Daniel Dennett (2013: 240), a philosopher:

1. Every mammal has a mammal for a mother.
2. If there have been any mammals at all, there have been only a finite number of mammals.
3. But if there has been even one mammal, then by (1), there has been an infinity of mammals, which contradicts (2), so there can't have been any mammals. It's a contradiction in terms.

Despite subsequent criticism, Dennett's statements illustrate the use of truncated logic in identifying implications. Specifically, Dennett did not state the rule by which the first assertion contradicts the second. Of course, the conventions of the English language point to an "infinity of mammals" as contradicting "only a finite number"; but it is the reliance on the conventions of a natural language that makes a mishmash of assertions in the sociological literature, so much that it is impossible to determine which components were deduced from others, let alone how.

In the case of Dennett's statements, we should note that they largely reflect an ancient tradition in philosophy, creating the impression that empirical observations are not really needed to assess empirical assertions; "reasoning" is sufficient. But to his credit, Dennett also took an extra-logical path, employing empirical evidence to conclude that some mammals did not have a mammalian mother. "If you go back far enough in the family tree of any mammal, you will eventually get to the *therapsids*, strange extinct bridge species between the reptiles and the mammals" (2013: 24).

A Predictive Problem for Science, Whatever the Substance

A predictive approach in assessing assertions assumes some stability in how the world works, but that assumption has been questioned. As a case in point, Lee Smolin (2013: xiv), a well-known theoretical physicist as well as a graduate faculty member of Toronto University's philosophy department, has stated: "Whatever is real in our universe is real in a moment of time, which is one of a succession of moments. The past was real but is no longer real." Surely the statement is an implicit denial of stability in events and things, perhaps to the point of even rejecting the very notion of categories (or classes) of kinds of events and things. But what else is implied?

In a related but distinct version of "the past is not the present" argument, Rupert Sheldrake (2012) has asked, what if the laws of nature change? To answer, he examined data related to possible variation in three assumed constants in physics—the universal gravitational constant, the fine-structure constant, and the speed of light. His controversial conclusion: there seems to be some variation. That conclusion raises follow-on questions. Do these variations arise simply from measurement error? If measured accurately, do these constants oscillate within fairly narrow limits? Are the fluctuations similar to weather patterns, regular but currently difficult to predict with precision? Are they simply chaotic? The questions suggest that nature may be more fickle than scientists suspect. If so, Sheldrake (2012: 93) concludes: "The implications of varying constants would be enormous. The course of nature would no longer seem blandly uniform; there would be fluctuations at the heart of physical reality."

Now the pursuit of knowledge gets interesting. Smolin's argument, along with Sheldrake's, generates doubts about the very idea of pursuing knowledge through predictive implications. The most obvious implication is that accurately predicting future findings from past findings is suspect at best, if at all defensible. For that reason alone, if Smolin's assertions are accepted, prospects of defensible conclusions from tests are minimal; and science as we know it would fall apart. Interesting is hardly a word for it.

Whatever the merits of the contentions made by Sheldrake and Smolin, it does not follow, as some constructivists and deconstructionists argue (e.g., Gergen 1994 and 1999) that knowledge is ours to make, not "out there" to be discovered. Evidence does matter. Even Sheldrake's paradoxical argument that "constants" may be "variable" is ostensibly based on evidence gathered, not ruminations while sitting in a rocking chair, smoking a cigar and sipping whiskey.

Although sociologists are nowhere close to discovering universal constants in social life, they can certainly continue adjusting predictions as generalizations are refined (e.g., Sewell, *et al.* 2001) and cease tolerating assertions that have no testable implications. However, if progress is to be made in these matters, conceptual clarity is as essential as it is currently lacking.

Illustrations of Conceptual Ambiguity

Virtually by definition, an ambiguous substantive word/term precludes sufficient empirical applicability, and ambiguity can be reduced only by the pursuit of conceptual clarity. In that connection, again consider Professor Joe Feagin's presidential address to the American Sociological Association in 2000 wherein (2001: 1–2) he gave "a few examples of troubling conditions currently being created or aggravated by modern capitalism," including "recurring economic disruption, exploitation, marginalization, and immiseration." Feagin issued a call to action, infused with provocative assertions about the detrimental impact of capitalism. However, we write *not* to dispute Feagin's indictment but, rather, to point out that he failed to formulate the kinds of definitions and generalizations that would permit a systematic empirical assessment of his argument.

Homosexual Genes

Less emotion but comparable ambiguity can be found in a question posed and answered in E.O. Wilson's *On Human Nature* (1978: 144), a book that promoted widespread adoption of sociobiology among social scientists, including anthropologists, sociologists, and psychologists (van den Berghe 1979; Lopreato 1984; Tooby and Cosmides 2005; Buss 2004). Prior to the widespread use of surrogate mothers and sperm banks, Wilson asked: "How can genes predisposing their carriers toward homosexuality spread through the population if homosexuals have no children? One answer is that their close relatives could have had more children as a result of their presence."

Wilson's answer is a study in ambiguities. First off, what are we to make of Wilson's "one answer"? Does it mean that the presence of homosexuals is really only a necessary condition for greater reproduction of their heterosexual relatives? Alternatively, Wilson might simply be enticing future research. Be that as it may, how could the mere *presence* of a close homosexual relative result in greater reproduction? At various places Wilson suggests that homosexuals somehow "help" or "assist" their heterosexual relatives, but those words can be interpreted in all too many different ways, with very different empirical implications. Then what are we to make of the "could have had more children"? Is the assertion no more than the identification of a "possibility"? If so, "fewer children" is another *possibility*.

Predictive Implications

In light of these ambiguities, it would be ludicrous to view Wilson's assertion as having any *testable* predictive implications. Worse, testable or not, the apparent predictive implications are obscure; and even the *least* obscure implication is extremely questionable. Wilson suggests that genes or genetic material "predisposing their carriers toward homosexuality" have been identified. This remains highly debatable, embedded as it currently is in the emerging field of epigenetics (e.g., Moore 2001 and Carey 2012).

Although Wilson did not point to any findings supporting his assertion, his writing on sociobiology has been highly influential. It is certainly not difficult to construct an *imagined* summary of relevant research that would support Wilson's assertion, to wit: "The survey found that respondents who identified themselves as a nephew or niece of at least one homosexual uncle or aunt reported more monetary gifts from those relatives than those reported by nephews or nieces of heterosexual uncles or aunts." Even with that imagined summary in hand, we would not be done. Additionally, we would need to know at a minimum whether close relatives of homosexuals reported having *more* children than reported by others in the same survey. Otherwise, how would these close relatives have had the essential *reproductive* advantage?

All of the foregoing attempts to fathom the predictive implications of Wilson's argument are debatable. Thus, Wilson's defenders can contend that homosexuals assist their close relatives by means *other than* monetary gifts, and Wilson himself suggested that contention by describing at length the *possible* social services of homosexuals in prehistoric human social units. The point is not just Wilson's inclination to substitute conjectures for evidence. He does not recognize that the perpetuation of the presumed homosexual genes must be occurring *now* as well as in some distant past. To be sure, observations on homosexuality in *contemporary* hunter/gathering populations would be less conjectural than appeals to the distant past, but limiting evidence to those populations would extremely diminish support of Wilson's argument.

The Wilson case is an instance of something all too common among biologists and physicists, even those who are recognized as a world authority (as in Wilson's case). In their fields they characteristically write as careful scientists, but in writing on human behavior they are commonly just as inept as sociologists when it comes to identifying the predictive implications of their arguments.

Recognition of the Limited Treatment

It is imperative to deny that the notion of an assertion suggests an exclusive concern with isolated generalizations (commonly identified as "propositions"). The point is not just that a generalization may be a component of a theory, either as a premise or a conclusion (theorem). More important, all premises and even the conclusions are assertions.

Mentioning theorems parenthetically is all the more important, for they are assertions deduced from some theory's premises. Hence, the importance of identifying predictive implications is accentuated in the case of theorems, both their deductions and contemplated tests. They are assertions in a real sense of being closer to the necessity of gathering data. The immediate qualification is that theorems are *directly* testable only relative to the "test proximity" of premises. Literally, only specific hypotheses derived from theorems or from isolated propositions can be tested directly.

To illustrate all too briefly, imagine a table in which there are two columns of values, *A* and *B*, where each pair of values pertains to one of several high school districts in Boston as of some particular year. Now suppose that the table was created to report a test of a hypothesis derived by some stipulated kind of logic from this truncated theorem: "Family stability is directly related to educational achievement." The wording of the hypothesis would be quite different, perhaps something like "There is a negative rank-order coefficient of correlation between the values in column *A* (e.g., percent of single-parent families in the district) and the *B* values (average test scores of the district students) in the table." The reported test of the hypothesis would consist of a "descriptive statement" as to the *actual* rank-order coefficient of correlation between the values in question (it could be negative or positive). Granted arguments about evidentiary relevance and the need for a test procedure that stipulates rules of derivation that involve epistemic statements (see Gibbs, 1994: 189–229 and 333–354), how could the hypothesis or the reported test be construed as ambiguous?

So a hypothesis is an assertion in the way of a prediction that can be declared confirmed or unconfirmed. However, the report of any association measure (an actual value) is itself an assertion infused with some uncertainty, for computational mistakes and even fraud are possible. Even the values in tables are tacit assertions, the claim being that the data and the computation of the values are in accordance with instructions as to how the theorem or proposition in question is to be tested (i.e., a stipulated test procedure). So a report of any test outcome comprises a series of assertions. No getting around it. That characterization poses a problem for any science, not just sociology. Nevertheless, lest this admission prompt philosophers to wring their hands and declare the game over, these problems have been present throughout the past five centuries and have hardly precluded achievements in science.

Unfortunately, our treatment of the importance of logic, empirical applicability, and conceptual clarity has been all to brief. Digestion of arguments in this chapter will be (we hope) assisted by further detail in subsequent chapters. For now, it is enough to underscore the importance of thinking of a scientific theory as nothing more or less than an organized collection of assertions in the form of generalizations, some of which have been derived from the others. A defense of this characterization of a scientific theory must await Chapter 5.

By way of conclusion, our argument in subsequent chapters is guided by the foundational belief that the foremost challenge posed for any scientific assertion is embodied in the question: what are the predictive implications of the assertion? Given this starting point, the ultimate challenge for scientists lies in another question: how can we state our assertions so as to enhance agreement as to the predictive implications of those assertions? Virtually all of the subsequent chapters focus on this question. We begin with what we mean when we talk about "science."

3

THE NOTION OF SCIENCE

Complexities and Problems

Why confront the reader with definitions of science? We might as well suggest spending the afternoon reading dictionaries; fun for some but not many. Nevertheless, justification is simple. Definitions more than matter; they are crucial in science, and all the more for sociology.

Yes, we frequently use words, terms, or phrases with little attention to definitions and more to related feelings and values. Some writers even go so far as to tacitly deny the importance of definitions, if not the possibility of formulating one. For instance, Walter Laquer wrote extensively on terrorism but denied (1977: 5) that the word can be defined comprehensively. Then, consider the question: are you an atheist, an agnostic, or do you believe in God? For some there will be an immediate and deeply personal response, filled with emotion, but few specifics: "Yes, God is a constant, powerful, and caring presence in my life." Others might reply with equal conviction though more specificity: "It depends. If by God you mean a supernatural being who is both omnipotent and benevolent, then I am an atheist. If God is omnipotent and benevolent, why are there ever-present natural disasters, such as hurricanes and tsunamis? If God is omnipotent and benevolent, why did he not strike Hitler dead before 1939 and avoid the death and suffering of some ten million persons?" Yet others might respond with less adamant resignation: "If you mean do I believe in a supernatural being who is omnipotent though indifferent to humanity's suffering, then I am agnostic." Needless to say, public opinion polls leave no room for such distinctions.

Should the importance of definitions still be doubted, contemplate three contentions. First, whether there is extraterrestrial life depends on the way life is defined, and there is an astonishing variety of contending definitions. Second, the very future of astrobiology (aka exobiology: the search for life beyond Earth) hinges appreciably on the definition of life. Third, even in the case of Earth alone, definitions are central for any "theory of life" and also for speculation about the origin of life. To repeat with added emphasis: definitions matter; they more than matter.

Some Initial Ground Clearing

Before we dive into definitions of science, we need to clear some ground. Many sociologists see themselves, often without acknowledging it, as social critics. Social criticism is an important activity, but it is not science. It is an exercise in value judgments.

Social critics may rely on data when they describe such things as the level of income inequality or the history of slavery. They may even make assertions about how the world works, but they do so with a view to promoting or preventing social change, not to identifying and verifying the predictive implications of their assertions. They are "closet ideologues," engaged in what may be worthy endeavors but not science.

This is not to say that applications of science in seeking valued outcomes should be avoided. Indeed, becoming social critics ourselves, no endeavor has been more important than science in shaping the nature and quality of human lives over the past three centuries. It informs, enriches, and threatens. Understanding the flows of electrons helps us store food, invent solar cells, and perfect microchips. Atomic bombs threaten our existence, while controlled nuclear fission provides a useful source of energy. Refining our knowledge of neural networks enhances our understanding of learning and memory. The field of epigenetics helps clarify the interaction between nature and nurture in the search to remedy birth defects. The mathematical models of fluid dynamics are critical to the design and efficiency of aircraft and cars, taking us from here to there and back. The theories and data related to the evolution of species and the sources of global warming inflame religious debates and stimulate political campaigns. And so it goes.

Clearly, value judgments and science intermix, but they are distinct endeavors. There are those who proclaim that such intermixing precludes value-free social science. To this we reply—right in one sense, wrong in another. Values influence the questions we choose to pursue and the applications we choose to develop, but we should not confuse the choice of questions with identifying and testing the predictive implications of assertions.

A Collection from Dictionaries and Encyclopedias

So, what is science? There are those who contend that no single definition can apply equally well to all sciences. Particular examples come from sociologists who rail against taking physical science as a model for inquiry. For example, Herbert Blumer (1969: 23) argued that the social and psychological sciences have been perniciously damaged in their attempts to see the empirical world in terms of investigative models developed in the physical sciences.

By contrast, Donald Black, in pursuit of what he takes to be "pure sociology," (Black 1995: 831, 847) has ventured:

> How, then, should my writings be evaluated? Scientifically ... If you wish to criticize my work, tell me you can predict and explain legal and related behavior better than I can. Tell me my work is not as testable as something else, tell me it is not as general as something else, tell me it is less elegant than something else, tell me that it has already been published, or just tell me it is wrong. Tell me something relevant to what I am trying to accomplish–something scientific.

While many naysayers do not flatly deny that sociology is, could be, or should be a science, they might as well. As for critics in the advanced sciences and in the philosophy of science, they may chortle at the very idea of sociology as a science. If so, they have confused criteria of science with scientific accomplishments. Now, we turn to the details.

Why not consider only definitions in the literature of science, the philosophy of science, and the history of science? The answer is interesting. Curiously, contributors to that literature rarely bother to formulate a careful definition of science, let alone an extensive conceptualization.[1] As for exceptions, the ostensible membership of contributors in opposing schools (e.g., positivists vs. anti-positivists, though those labels are virtually meaningless) suggests that their definitions reflect what they think science *should be* more than what they think of as "actual science."

Some Choices

The principal alternatives to definitions in the philosophy/science literature are those in dictionaries and encyclopedias. However, a collection of definitions from these sources can be a random sample only from some arbitrary and limited category (e.g., all dictionaries in English published after 1999). What follows reflects selection directed by accessibility—one personal and the others in the holdings of a large public library. Such selection, while not random, promises some degree of broad coverage, in that no definition was selected or excluded on preordained criteria as to what science is or is not, should or should not be.

The search of dictionaries and encyclopedias culminated with eleven definitions, all of which are in Appendix I. Even if the reader could suffer through it, the eleven definitions need not be read, for each is an instance of one of four illustrative types.

The Type I Definitions: The Irrelevant

As described subsequently, a thoughtful definition of science is likely to reflect the resolution or avoidance of a conceptual issue or problem. However, some definitions appear indifferent to issues and problems, or skate over them so quickly that they are irrelevant for our purposes. One instance among the eleven in Appendix I is that of Robert Barnhart (1988: 968), who sees science as "The modern restricted sense of a branch of learning based on observation and tested truths, arranged in an orderly system, is first recorded in English in Israel Watt's *Logic* (1725) ..."

For reasons described later, Barnhart's definition could be identified as an instance of the "distinctive activity" type by virtue of the word "observation"; but obviously the activities of scientists qua scientists are not limited to observing, and billions of "observers" are not scientists. Moreover, virtually all scientists write reports, perhaps with a view to publication. For that matter, observation and classification are often linked, but they are distinct scientific activities. As for Barnhart's "tested truths," they could be construed as products of science but surely only some of many. For that matter, a multitude of sophisticated scientists appear wary of the very notion of "truth," and understandably so.

Although the judgment of it may be all too harsh, Barnhart's definition is simply irrelevant when it comes to major issues or problems in defining science. An even harsher judgment: in being irrelevant, the definition is worse than wrong.

The Type II Definitions: A Distinctive Kind of Product

To define science effectively we need to address the distinctive nature of both "activities" and "products." Although some activity is necessary for any human product, many definitions of science appear much more concerned with *kind* of product than with related activity. Fortunately, definitions of science commonly agree as to the general kind of product. As one instance, the word "knowledge" appears in nine of the eleven definitions in Appendix I.

Of all eleven definitions of science, Frederick Mish (2003: 1112) has formulated the most obvious instance of the Type II definition:

> 1: the state of knowing: knowledge as distinguished from ignorance or misunderstanding 2 a: a department of systematized knowledge as an object of study <the ~ of theology> b: something (as a sport or technique) that may be studied or learned like systematized knowledge <have it down to a ~> 3 a: knowledge or a system of knowledge covering general truths or the operation of general laws esp. as obtained and tested through scientific method b: such knowledge or such a system of knowledge concerned with the physical world and its phenomena: *natural science* 4: a system or method reconciling practical ends with scientific laws <cooking is both a ~ and an art> 5 *cap: CHRISTIAN SCIENCE.*

Principal Merits

The contention that "knowledge" is the primary if not exclusive product of science is hardly radical, but it diverges sharply from viewing science in material terms. That "material view" is consistent with something that few would dispute. For centuries and especially after 1800, science has been associated one way or another with all manner of inventions and perfections of various machines, instruments, or energy sources (see, especially, the National Geographic, 2010). The use of electricity may be the leading case, but the atom bomb and subsequent development of atomic energy are surely contenders. Nonetheless, it would be ridiculous to characterize all products of science—theories and conceptualizations in particular—as "material," let alone useful or even beneficial; and the definitions in Appendix I avoid that characterization.

Another merit of Mish's definition is the suggestion that "scientific knowledge" is predominantly organized in some sense or another, as opposed to isolated bits and pieces in the way of a list of commonly accepted "facts." The suggestion is made by Mish's frequent use of the word "system" or "systematized," but he does not recognize that those words apply to such things as a mileage chart for U.S. cities and numerous other kinds of "knowledge" that are of limited scientific relevance. Moreover, scientific knowledge is most conspicuously organized in the form of commonly accepted theories, but the word theory or theories is not in Mish's definitions or, curiously, any in Appendix I.

Problems and Issues

Even though not always made explicit, when speakers or writers use the word "knowledge" in connection with science, they frequently appear to mean beliefs or statements that are "true" in one sense or another; and that is the case for all definitions of science in Appendix I. When truth claims are made, a host of problems appear—the *dilemma of induction* being foremost. These cannot be circumvented by logomachy, simply switching from "true" to "valid," "verifiable," "tested," or "confirmed." Whatever the terminology, the problem is an ancient one that reduces to this question: how can the truth of a belief or statement be demonstrated?

Beyond the point that predictive implications are necessarily involved, answers to "the truth question" are debatable even when the subject is a particular event or thing and not a general claim concerning how the world works. Thus, it appears that the identity of President Kennedy's assassin or assassins will be debated until the cows come home, as will assertions about causes of America's Civil War. The problem is exacerbated when scientific knowledge is extended to beliefs or statements in the form of generalizations about broadly construed categories of events or things, especially *infinite categories*—those that have no spatial and/or temporal limits (e.g., all assassinations, all civil wars).

For some, the seeking of indisputable truth is a fool's errand, and the seeking extends even to the notion of "falsifiability." For example, Philip Kitcher (1985: 58–61) rejects Karl Popper's "falsifiability criterion" of science, one rationale being the assertion that "In one obvious sense, evolutionary theory, sociobiology, and all major scientific theories are unfalsifiable."

The truth of generalizations has been debated (especially in light of David Hume's position on the matter) long before its rejuvenation by Popper's foundational contention that generalizations in the form "All *X*'s are *Y*'s" cannot be verified. And his critics, such as Kitcher (1985), are deluded in assuming that rejection of the "falsifiability principle" will negate Popper's contention. However, Popper notwithstanding and many definitions of science to the contrary, an *untested* set of statements *may be* justifiably identified as a scientific theory. It is even difficult to see why a widely rejected theory ceases to be a theory.

Of course, the issue can be simply ignored by declaring that scientific knowledge comprises distinctive kinds of generalizations and theories, and the stipulation of the kind need make no reference to truth or even testability. Such declaration might differentiate science from other human endeavors, conventional history and journalism in particular. However, if scientific generalizations or theories are distinctive in that they apply to a wide range of situations or even have an infinite quality (no spatial and/or temporal limits), there must be serious doubts about the standing of various widely accepted scientific theories. They include the theory of continental drift or plate tectonics and even the Copernican theory.

The foregoing should not be construed as a rejection of "objective reality" or acceptance of any other related tenet of deconstructionism, postmodernism, *ad nauseam*. Specifically, there are all manner of beliefs or statements about *particulars* that only the deranged would question. But such beliefs or statements are scarcely "scientific" knowledge. Again, think of a mileage chart for U.S. cities.

The Type III Definitions: Science as a Type of Activity

Of the eleven definitions of science in Appendix I, the word "activity" exceeds "knowledge" only in Morris's definition (1969: 1162), but in all eleven definitions the word "activity" or "activities" appears at least once in only four, and there are only eight appearances, with William Morris's definition accounting for four of them. Specifically, Morris (1969: 1162) sees science as:

> 1. The observation, identification, description, experimental investigation, and theoretical explanation of natural phenomena. 2. Such activity restricted to a class of natural phenomena. 3. Such activity applied to any class of phenomena. 4. Any methodological activity, discipline, or study. 5. Any activity that appears to require study and method. 6. Knowledge; especially, knowledge gained through experience.

Principal Merits

Such a definition is all the more strategic because whatever the products of science, knowledge or otherwise, they are obviously the result of some human activity. However, the reverse does not hold; an activity does not necessarily result in a product. Scientists and artists can readily testify that hours of activity have been wasted in that there was no desired product, nor even any prospect.

The inclusiveness of Morris's definition is much greater than appearance may indicate. This is especially the case when "theoretical explanation" is understood to include both the formulation and the promulgation of theories. As pointed out earlier, that merit is all the more noteworthy in that the eleven definitions suggest a remarkable omission of explicit concern with theories. Of course, Morris's "theoretical explanation" may encompass a wide variety of meanings; and if unclear, it is not peculiar to Morris. Rather, it is an intractable problem in assessing definitions, whatever the *definiendum*, but it is intensified in the case of "science" by the common use of such words as knowledge, activity, activities, method, and methods.

Morris's definition and any other Type III instance is likely to be most appreciated by historians of science, especially those with a special interest in the origins of science, and that appreciation is one of the definition's merits. However, any idea on that subject must be conjectural, though it taxes credulity to argue that science commenced from the get go with the application of some particular method, let alone a theory. It is far more plausible to assume that science commenced with observations. The kinds are debatable, but "observing" is certainly an activity, though of no scientific consequence unless it leads to a generalization or the questioning of one.

Still another merit of Morris's definition is that it goes further than many of the other eleven when it comes to the identification of specific kinds of activities deemed to be science. Yet the definition ignores various "communication activities," such as writing reports of research and perhaps attempts at publication. However, it is extremely doubtful, if feasible at all, to frame a definition of science that recognizes all supposed kinds of *distinctive* scientific activities, even general kinds.

Problems and Issues

As indicated by Morris's definition, numerous human activities—observing, describing, classifying, and measuring—are commonly mentioned in connection with science. Yet it would be ridiculous to argue that those activities are peculiar to science. After all, everyone observes, journalists describe, historians classify, and tailors measure.

The problem surely cannot be resolved by maintaining that an activity is distinctively scientific if and only if engaged in exclusively by scientists.[2] Even if such activities could be identified, the problem of defining science would become one of defining "scientist." The latter may well pose even greater problems, commencing with recognition that much of the behavior of scientists is scarcely scientific, as their associates, spouses in particular, will readily testify.

Type IV Definitions: Science as the Application of Distinctive Methods

Commentaries by philosophers and scientists on science are replete with reference to the "scientific method," as though some method is peculiar to science. Yet the word "method" or "methods" appears at *least once* in only six of the eleven definitions in Appendix I, with a total of just seven appearances. By contrast, "knowledge" appears at least once in nine of the eleven definitions, with a total of twenty-six appearances. Of the six definitions in which "method" or "methods" appears at least once, they appear more than once only in Mish's definition, which was quoted in connection with "knowledge" and need not be repeated.

Problems and Issues

The word "method" or "methods" does not appear even once in five of the eleven definitions of science. That remarkable lacuna could largely reflect the questionable reliance on "word counts" in assessing definitions, whatever the *definiendum*. Doubts about that reliance are especially justified in light of the inclination of scientists (e.g., Cromer 1993: 144) to stress the importance of "mathematics" and "experimentation" rather than speak of method or methods in some more general sense.

The less obvious problem is that the very notion of "method" cannot be readily separated from the notion of "activity," especially when it comes to the application of methods. So there is a related semantics question, and it is difficult. Observing, describing, classifying, and measuring are surely activities, but are they also "methods"? An affirmative answer creates doubts as to what constitutes a truly *distinctive* "scientific method." Again, everyone observes, journalists describe, historians classify, and tailors measure.

Apart from this semantics problem, in the literature there are many flat denials that scientists employ some distinctive method, and that denial sometimes comes from prominent but controversial figures in the history or philosophy of science (e.g., Feyerabend 1975). Nor is the denial puzzling, especially as regards experimentation and the use of mathematics. In light of the history of astronomy, geology, and all human sciences possibly save psychology, how can experiments be an essential scientific method?[3] Furthermore, if mathematics is an essential method in science, what of Darwin's *Origin* (1996 [1859])? Try to find equations in it!

The Merits Question

For reasons previously suggested, it is difficult to identify any particular merit of a Type IV definition of science. The problem commences with the real possibility that there can be no clear-cut distinction between "method" and "activity," especially given that (1) the very idea of a "method" apart from its potential application makes little sense and (2) the actual application of anything is an activity. Those links force recognition again that Type IV definitions do not identify distinctive kinds of scientific activities.

As indicated by our proposed alternative definition of science coming up in Chapter 4, there is a kind of activity that scientists engage in frequently. It is the formulation and assessment of theories or isolated generalizations, both about empirical relations. Yes, philosophers also formulate and assess theories or generalizations, but whether those "products" pertain to empirical relations is quite another question. Even when the products appear "empirical," philosophers certainly do not employ a particular scientific criterion to assess those products. The nature of that criterion is dealt with in Chapter 4, but even at this point note that the very notion of a distinctive criterion is a conspicuous lacuna in the eleven definitions of science in Appendix I.

The coupling of "product" and "activity" in defining science can be justified without explicit reference to some method. Briefly, scientific knowledge is a product of a distinctive activity, distinctive in being the application of a particular criterion or criteria. The general idea is simple enough. Whatever the form of statements—generalizations or otherwise—they are not scientific unless certain criteria are satisfied. Unfortunately, even when commentaries on science do suggest "distinctive scientific criteria," they tend to be extremely ambiguous as to the nature of the criteria, and even more so with the regard to "satisfying a criterion." Further, for reasons indicated earlier, the ambiguity is not resolved by appeals to a scientific method.

With these critiques in hand, what do we propose as an alternative? Coming up in the next chapter.

Notes

1 As an illustrative case, Geoffrey Gorham, a philosopher of science, wrote an entire chapter ("Definition of Science," 2009: 27–52) without even coming close to defining science, though offering a commendable survey of problems. Paul Thagard (2012) topped Gorham in writing *The Cognitive Science of Science* without even mentioning definitions of science. Then read chapter 14 ("What Is Science?") in a book authored by a physicist (Omnes 1999) and try to find an explicit definition of science. Or consider Lee Smolin, a theoretical physicist, who wrote a chapter (2006: 289–307) titled "What Is Science?" without posing a definition other than in the last sentence, quoting Richard Feynman: "Science is the organized skepticism in the reliability of expert opinion." The quote is far from just "clever" (see Cromer [1993: 143–144]). The ultimate case: Thomas Kuhn (1996) wrote what may be the most cited book in the history/philosophy of science without defining science, despite a section in the "Postscript" (1996: 207–210) designated as "The Nature of Science."

2 Throughout this book, unless explicitly indicated otherwise, the words "science," "scientific" and "scientist" are used in the putative sense, though recognizing that what is "general" or "common" can be disputable.

3 In the *Macmillan Encyclopedia of Physics* (Rigden 1996: 1412) Richard Feynman, a giant in the history of American physics, is quoted thus: "The principle of science, the definition, almost, is the following: *The test of all knowledge is experiment*. Experiment is the *sole judge* of scientific truth." Incidentally, the publication is one of several encyclopedias and dictionaries in which there is an entry for "scientific method" but not for "science."

4

A CONCEPTUALIZATION OF SCIENCE

This chapter continues the journey into the conceptual territory. You may find yourself asking: "Are we there yet?" Not yet, but we are getting closer. Let us begin with a definition of science, though our conceptualization includes much more.

Science is here defined in terms of what its practitioners *do*. Science consists of (1) activities believed by those who participate in or control the activities as (2) conducive to the formulation of assertions about the characteristics of entities and (3) assessments of those assertions in light of their predictive power relative to that of contenders. If you nurse misgivings about the notion of "scientist," observe the word does not appear in the definition.

We offer this definition with the intent to cover science from its very awakening to whatever the future might hold. Additionally, the desire is to reveal as precisely as possible what we have in mind, believing it feckless to assume that one can have intelligible views of science that do not at least imply a definition of it.

In offering the three-part delineation of science we hope to minimize inscrutable observations on and esoteric arguments about science. We hope also to separate "doing science" from other pursuits and thereby avoid starkly broad formulations, such as Wallace Matson's depiction of science (2011: 208) as "The activity of using reasonable procedures to find out what things (including ourselves) are really like; and the product of that activity." Given Matson's all-encompassing definition, we admit to a little amusement on reading the further claim (Matson 2011: 74) that "Science began only once in human history: in the city of Miletus in Ionia, early in the sixth century B.C. And it was the creation of one man, Thales …"

The Three Components of Science

Science can be thought of as activities (e.g., observations, experiments) or as products (e.g., hypotheses, theories). Our definition includes both, assuming products require activities, along with assessments of assertions about characteristics of entities in light of the relative predictive power of those assertions.

Activities

The first component of our science definition is "activities." An activity may be *overt* behavior (e.g., moving about in some animal's habitat), *internal* behavior (e.g., thinking about features of the animal), or some *combination* (e.g., recording observations). So "activities" encompasses a very wide range of behavior. Indeed, it is in this realm of broadly defined activities that postmodernists ground their incessant claim that truth is socially created (e.g., Gergen 1999). More of this later. Less obvious, the activity component differentiates our definition of science from one pertaining exclusively to products, a common practice. In our view science is both process and outcome. Knowledge is the most common product, but it is never independent of some "producing activity." Moreover, tradition notwithstanding, putative knowledge is actually a collection of assertions.

Much can be learned about "doing science" by reading biographies, autobiographies, and memoirs of scientists.[1] One engaging illustration of the interlaced activities of scientists is found in relation to Charles Darwin's *Origin of the Species*, including its link with the work of theologian, economist, and demographer Thomas Malthus. As the eighteenth century drew to a close, Malthus published a brief pamphlet: *An Essay on the Principle of Population as it Affects the Future Improvement of Society*. Malthus's assertions were simple enough. Humans need food. Humans enjoy sex. Sex produces offspring. Left unchecked, the population will grow geometrically, while food supplies will grow arithmetically. Disease and famine will follow as sure as night follows day. Although Malthus' theory is questionable for various reasons, it remains very influential; and it is likely to become even more influential in light of global warming.

Thomas Malthus died in 1834, and the second edition of his essay was published two years later. Around this time, a young Charles Darwin, not yet 30 years old, had just returned from his circumnavigation of the globe on the *Beagle*. He intended to pursue what he referred to as his systematic inquiry. According to Darwin's own account (1876: 47), Malthus's brief essay provided important grounding for his own work:

> I happened to read for amusement Malthus on Population, and being well prepared to appreciate the struggle for existence which everywhere goes on from long-continued observation of the habits of animals and plants, it at once struck me that under these circumstances favourable variations would tend to be preserved, and unfavourable ones to be destroyed. The results of this would be the formation of a new species. Here, then I had at last got a theory by which to work.

It would be some twenty years, involving extensive communication among colleagues—notably William Wells, Alfred Wallace, and John Herschel—before the *Origin of the Species* was published, but a distinctive theory of evolution had begun.

Assertions

Some activities are designed to produce knowledge, which brings us to the second component of our science definition. Putative knowledge is an express or implied answer to a question, stated or unstated. Thus, obviously, "Abe Lincoln was born in Kentucky," answers the question "Where was Abe Lincoln born?" Lest this illustration appear silly, statements about instances ("factual" or "instantiative" statements) are staples in any science; and each one could be an answer to a question. Would answering the "Lincoln question" be a scientific activity, or simply an instance of satisfying general curiosity? Like so many questions about scientific activity, the answer depends on context. In this particular case, if an answer to the Lincoln question was intended to enter into a test of, say, a generalization or theory about governing elites, the answer would qualify as a scientific activity.

The larger and more important point: again judging by historical narratives, biographies, autobiographies, memoires, and conversations with self-identified scientists, there is no "logic of discovery." The history of science is filled with a multiplicity of paths taken to generate questions and the activities engaged in pursuit of answers. Fortunately, this is not a fatal problem, which is to say that how one arrives at a generalization is irrelevant as regards its predictive power, the third component of our science definition.

Even so, it is important to note that multiple and sometimes winding paths to a theory may be important, if only because "getting the question right" is central to the enterprise. The distinguished sociologist Robert K. Merton (1910–2003) placed the activity of identifying theory-relevant questions center stage, but that emphasis has declined. It was replaced with frequent contentions of postmodernists that the questions pursued by scientists are determined by interests *external* to science.[2] According to opponents of the traditional view of science, the activities of scientists and the products reflect the interests of some social class, those of a more narrowly defined interest group (e.g., feminists), those determined by political hot topics (e.g., the impact of same-sex relations on raising children), or currently the politics swirling around global warming.

Interestingly, in expressing their arguments about science the "externalists" rarely describe *actual specific attempts* by outsiders to control scientific activities. Surely there are instances, but frequency and success are debatable in the extreme. In that connection, it is important to draw a distinction between (1) external influences on the substantive focus of the questions being asked (e.g., etiology of cancer, extent of global warming, correlates of crime) and (2) external influences on the answers. In the latter case, corrupted science results. As to the former, surely a focus in science on the relative predictive power of assertions provides some shield from external influence. When science is involved, predictive power should be the focus, whatever the topic being investigated, and whatever the motivation behind the questions being asked.

Without doubt, in selecting questions to pursue, various interests will always be promoted, whether by scientists or others. So when the Director of the National Science Foundation appears before a legislative committee, her/his plea for funds to support research on some science-based question may appear to be political rhetoric, an exercise in self-interest, or simply obligatory. But any such interpretation surely stops short of a denial that (1) the Director wants to promote scientific activity and (2) one way or another he/she has attempted to control those activities. "Promotion" is one of many words suggesting a "control attempt," and our conceptualization of the latter has been published elsewhere (Gibbs 2008: 3–49).

If only because of the concern with the relative predictive power of their assertions, scientists need to obsessively avoid corrupted answers to questions; and they could simply grant that only rarely is there a clear-cut scientific justification for pursuing one set of questions rather than another. Corruptive influence on answers may be blatant, as when a tobacco corporation pays a researcher to slant if not falsify research findings, but subtle and undetected bias can enter into research as well, even when gathering data.

Consider the Human Relations Area Files (HRAF), an internationally recognized data set created at Yale University in 1949. The goal was ambitious—to establish a source of information for comparative global studies of what are conventionally known as "cultures," past and present. To that end, ethnographic and archaeological findings were gathered and organized. This source of comparative data was designed and did become a treasure trove for the social sciences. Unfortunately, there was and remains at least one serious limitation.

Success of the enterprise required a representative sample of cultures. Some twenty years after HRAF was created, Raoul Naroll (1967, 1968) emphasized what he and colleagues had worried over for years. Even ignoring the nebulous meaning of "a culture," the files were not a random collection of entities (tribes, kingdoms, etc.); hence, generalizations from the files to all cultures are inherently questionable. The interests of ethnographers (some really proselytizing missionaries), national politics, and ease of safe travel—to mention only three selective factors—meant that the HRAF were likely to disproportionately include entities that are or were distinctively colorful, large, and near transcontinental routes of communication. In subsequent years, remedial steps were taken by HRAF personnel; but in the absence of random sampling procedures *ab initio*, the extent of bias in HRAF data remains a crucial problem in both formulating and testing generalizations.[3]

This brings us to a more general point. Ideally, answers to scientific questions are implied by accepted theories, but in recent decades increasingly little attention is given in sociology to assessments of theories strictly in light of tests. Instead, we have "interpretations" or "critiques." The practitioners are prone to treat tests, if at all, as controversial, and criteria for assessing theories are rarely stated. Rather, critiques of a theory commonly appear preoccupied with attributing dubious assumptions to it, without mention of tests or even the need for them.

Such assessments have become a cottage industry in sociology, where one can search leading journals endlessly without finding a report strictly limited to a test of some theory. Instead, we commonly find assessments in terms of the perspective (Marxist, functionalist, feminist, etc.) that supposedly gave rise to the theory.[4] If the assessor takes a dim view of that "generative perspective," then so much the worse for the theory.

At this point it is important to note that our definition of science makes no reference to "theories," a subject treated at length in Chapter 5. Yet we do recognize that any theory answers *at least one* question, though both the question and the answer may be only implied by the theory. Indeed, Newton's theory came to be celebrated precisely because it implies, as Newton himself recognized, answers to various questions, about tides, planetary motions, etc. Einstein's theory came to *supplement* Newton's because it answers even more questions.

A final note on activities and theory construction: To our knowledge no extant science commenced with a theory (virtually inconceivable), but all commenced with activities of some kind. Even in contemporary sciences, the practitioners often pursue questions that have *no* connection with any particular theory (e.g., the size of some animal's range or the amount of income inequality in the U.S. as of some particular year). The point is made to counter the common but untestable assertion that all data, questions, observations, etc., are "theory laden." The response, beyond calling for the identification of a specific theory in each case, should be: surely not the first one!

What has been said about theories extends a fortiori to "explanations." An explanation even more obviously answers a "question." Less recognized, defining "explanation" entails very difficult problems and a variety of serious issues.[5] So, although various "scientific activities" may appear to have some connection with explanations, that connection should be avoided in defining science unless one has a real appetite for insoluble problems and irresolvable issues.

Before we move to the third element of our definition, consider this: critics are likely to insist that products of scientific activities pertain to "causes," not simply "characteristics." In reply, ever since David Hume the notion of causation has been a thicket of problems and issues that currently appear beyond resolution.[6] That thicket can be circumvented by recognizing that whatever the alleged causes or consequences of a thing or event, they must be alleged correlates of it. As for the bromide "correlation is not proof of causation," it suggests uncritical acceptance of the notion of causation and consensus as to causal evidence, even though consensus has not been the case in philosophy or science for centuries. If in doubt, read Bertrand Russell and Ernst Mach on causation. The bromide also ignores the variety of types of space-time relations (discussed subsequently), some of which may be much more relevant than others in making causal inferences.

So the exclusion of "explanation" and "causation" from our definition of science has a rationale, one best stated as a question. In defining any kind of human activity, why include words that have proven conducive to insoluble problems and unresolvable issues as regards interpretation? To extensively paraphrase Woody Allen on sex without love, defining science without mention of explanation or causation is bound to appear bizarre; but as bizarre things go, it is better than promoting hopelessly ambiguous and/or controversial notions.

Assessment of Predictive Power

We come now to the third and final component of our definition of science, which reduces to a declaration: if an answer to an empirical question is *not* susceptible to a direct or indirect assessment in terms of predictive power, it is *not* science. We freely grant that all components of our definition are disputable, and this is especially the case for this third component. Indeed, we predict, based on conversations we have had and tomes we have read, that a herd of philosophers and even some scientists will object.[7]

Two Logics

Throughout the history of the philosophy of science a debate has raged over two contending logics for evaluating theories, a logic of discovery versus a logic of justification. Proponents of the former argue that steps taken in the formulation of a theory are strategic as regards acceptance/rejection.[8] The issue extends to aesthetic considerations, including "elegance" (Glynn 2010) and even "mathematical reasoning" (Lindley 1993). Perhaps in large part because of the scope of the issues, proponents of "discovery logic" have not realized anything like agreement on an unambiguous and more informative description of that logic than we have given (*supra*).

Our objection is that how one arrives at a theory has no necessary bearing on the accuracy of predictions derived from it. The theory may come as the end product of numerous observations (i.e., crude induction, aka "dustbowl empiricism"), more recently exemplified by what is called "data mining" (Zaki and Meira 2014). Alternatively, it may have leaped to the theorist's mind via pure intuition, perhaps while intoxicated or waking up with a remembered portion of a dream, or it may have emerged as a product of pondering over some puzzle. It really makes no difference as regards the theory's predictive power. Statements by the theorist as to the theory's provenance may make it more plausible, and perhaps enhance the probability of it being published or tested if published. Nevertheless, when it comes to a logic of discovery, it has no necessary bearing on test outcomes.

By contrast, defenders of a "logic of justification" contend that there can be criteria for assessing a theory that have nothing to do with its origin or even steps in its formulation. Those defenders would grant that some kind of logic is necessarily involved in the deduction of theorems (or conclusions) from premises as well as deductions of predictions (hypotheses) from the theorems. The defenders might also concede that theorems are never demonstrably unique, though rightly protest that it is impossible to show that the theorems can be deduced only from the premises in question and thereby avoid the fallacy of affirming the consequent. In any case, deductions do not corroborate theorems. Specifically, with possibly some recalcitrant economists to the contrary, there is agreement that theorems are not true because the premises are *taken* as true.

The immediate problem with the "logic of justification" is that its advocates diverge in answering this question: ignoring a theory's provenance, what justifies its acceptance or rejection? One answer is "correspondence with reality," another "agreement with facts," and still another "consistency with test outcomes." The answer offered here: a theory or generalization should be judged in terms of its predictive power relative to contenders.

The only obvious alternative to a *relative assessment* of a theory or generalization is some absolute standard. In the *qualitative* case, the criterion must be either "no known exceptions" or, less stringent and far more realistic, a certain minimum percentage. Unfortunately, any percentage criterion is arbitrary, and in the case of *quantitative* theories or generalizations it would be grossly unrealistic to demand some particular *minimum* association between the variables, much less some particular consistent degree of association. The most unrealistic demand would be for consistently accurate predictions of precise values predicted from equations, particularly on recognition of varying degrees of uncertainty in any prediction, given measurement error and the stochastic nature of many empirical associations (not just in human phenomena). An absolute criterion is appropriate in the case of a purported "scientific law," but we doubt the very idea of invariant associations, let alone maximum associations, and especially in the social sciences.

In the absence of an *indisputable* contender, a theory deserves special consideration if tests indicate its predictive accuracy clearly exceeds chance (however reckoned). The only complication is the possibility of a *logical* relation in that, for example, theory *X* implies predictions about the *total* crime rates of cities, while theory *Y* enables predictions about the robbery rates of cities.[9] Appearances to the contrary, the two would *not* be contenders, all the more if they pertain to different kinds of entities (e.g., cities vs. age groups) and/or *only one* pertains to *change* in crime rates. Of course, when two theories are contenders, the scene is always set for a third theory—perhaps a synthesis—that exceeds both contenders as regards predictive power.

With all this said, it bears repeating throughout this book that "prediction" is not synonymous with forecasts, extrapolations, or prophecies, no matter how plausible. Forecasting, extrapolations, and prophecies do not require any generalization or theory; just "a hunch" will do.[10] Moreover, if a theory comprises generalizations about infinite categories of phenomena, it necessarily pertains to the past as well as the future; and, obviously, there can be no test based on data for circumstances yet to be realized.

If one insists that "predictions" always pertain to the future, what of evidence pertaining to the Big Bang theory or a theory about the French Revolution? What the prefix "pre" in predictions refers to is what we are likely to find once data are collected or observations/measures made, not to the occurrence of the phenomenon in question.

Finally, our definition of science is deliberately very broad but not to the extent of even suggesting that virtually all of human activity is "science." We featherless bipeds often ask or answer all manner of questions without the slightest thought about predictive implications, or contending answers for that matter.

The Seven Dimensions of Predictive Power

To say it again, in light of our three-part definition of science, all manner of particular "human questions and answers" do not qualify. The immediate reason is that they have little if nothing to do with predictive power, a notion that can best be described in terms of seven dimensions.

Testability

Of the seven dimensions of predictive power, testability is the most difficult to describe and poses the most serious issues/problems, all the more if one takes the philosophy of science seriously (see Sober [1999] on "testability"). However, it is refreshing to say that there are no insurmountable problems when describing a test of an assertion in *general terms*. Step 1: Identify some statement as an assertion worthy of consideration. Step 2: Make statements that link words or terms in the assertion with data or observations. Step 3: Formulate or adopt rules for deducing a prediction as an implication of the initial statement (the assertion) and the "linking statements." Step 4: Formulate or adopt a conventional rule for declaring the assertion (Step 1) as being consistent or inconsistent with the implicative statement (Step 3). Step 5: In accordance with Step 4, declare the two statements as being consistent or inconsistent, though perhaps as a matter of degree or amount. Step 6: Consistent with the public character of science, attempt to publish the Step 5 declaration or otherwise make it known to colleagues. In short, assert an idea, link that idea to data, observe, decide, and publish (or, alas, try to, and bear in mind that "publish and perish" is a distinct but seldom recognized possibility).

Should it appear that the foregoing description of a test is not needed, merely stating the obvious while preaching to the choir: Nonsense! You can easily crash your computer searching the recent sociological literature for a publication in which the author sets forth the test procedure and defends the predictive power criterion. Yes, there are instances where a sociologist warrants praise for her/his candor by promoting what is an alternative criterion, such as "intellectual satisfaction" (e.g., Charles Tittle [1996: 347]). What you will find in virtually all cases are not alternatives to predictive power; rather, profound silence on this fundamental issue.

As for physics, Steven Weinberg writes (2009) compellingly of "elegance" as a criterion. However, in that connection advocates of string theory (sometimes referred to as the theory of everything) have developed far-reaching ideas; for example, the possibility of multiple universes. Those ideas may or may not appear elegant. Even so, indifference to testability remains an art form for string theorists, at least currently. Intellectual satisfaction and elegance are certainly valued, as are metaphors involving the heuristic implications of interacting strings, but they have little to do with tests.

Only a rash individual will deny that tests of generalizations and theories are a major scientific activity, but in some philosophical circles the very notion of a test is tacitly questioned (again, Sober 1999), as it presumes a distinction between synthetic statements (e.g., "The Statue of Liberty is on an island.") and analytic statements, the latter true by virtue of the constituent words (e.g., all triangles have three sides). For W.V. Quine (1951 and 1960) and his "two dogmas of empiricism," the analytic/synthetic distinction should be replaced by what he calls the indeterminacy of translation.

Although Quine's indeterminacy of translation has become a staple of the philosophy of science, the issue warrants further attention if only because some philosophers of science strive to create the impression that science is haunted by truly crippling problems. These claims are generally made without confronting this crucial question: if the problem at hand is truly crippling, how has science progressed over many centuries? Similarly, it is difficult to deny the success of science without ignoring the increase in the number of accepted theories, as well as their expanding range and scope. If the word "theories" grates, substitute "generalizations" or "accepted explanations."

With this said, we quickly note that Quine and others have identified an important issue in that (1) ambiguities abound when we move from assertions to gathering data for assessments and (2) only the incorrigibly naïve will deny that in some instances the distinction between synthetic and analytic statements is extremely debatable. An example that comes to mind is the Darwinian generalization that only "fit" organisms tend to survive and reproduce.

The suggestion is not to deny problems with the analytic/synthetic distinction. To the contrary, Quine's argument has rightly received close attention. In the immediate context, contemplating tests of an assertion should commence with this question: is the statement true or false by virtue of the very meaning of its constituent language units (words, terms, or phrases) or syntax? Tests are not possible without examining those meanings, and the outcome inevitably depends on perceptions of some conventional language and/or acceptance of special definitions, if any, accompanying the assertion at hand. If there is an issue, it is settled one way or another by *effective* consensus in a particular field (e.g., all or virtually all "geologists"). The implied distinction between absolute and effective consensus means that science's adversarial character is reflected even in "language issues," but debates—whatever the issue—are far from crippling. Instead, they are at the very core of scientific progress, a point that undermines various arguments of philosophers about science.

Judgments of a generalization's predictive accuracy will forever remain conjectural unless based on actual tests, not impressions, assumed values of variables, or meta-theoretical presuppositions. We have seen an example of indifference to testability flowing from a Marxist perspective in the 2000 presidential address to the American Sociological Association. There are many more, several of which come from Talcott Parsons. He wrote (1951: 15): "Culture, that is on the one hand the product of, and on the other hand a determinant of, systems of human social interaction." The assertion may seem reasonable enough, but you might want to sink your teeth into it and then devise a defensible test. It will not be easy.

So the question: what is a testable statement? A statement is testable, at least in principle, only if synthetic; and defensible tests of a synthetic statement are possible only to the extent the statement's constituent words and terms are empirically applicable. A word, term, definition, formula, method, technique, or procedure is empirically applicable *to the extent that* investigators acting independently (1) indicate that it is intelligible, (2) report that its use is feasible when identifying or describing events and things, and (3) agree when applying it to components of an assertion (e.g., the annual official burglary rate of some particular city as of a particular year).

A test of an assertion requires that the assertion be both comprehensible and comprise empirically applicable words/terms that refer to real events or things. This is not always as straightforward as it appears. Consider simulations. They involve assumptions and conclusions, both of which can be totally devoid of contact with "real world" data. Equations are proposed, alternative processes constructed, and super computers employed to simulate weather patterns, earthquakes, or the structure of molecules. These simulations may then be compared with actual outcomes, but until so compared they remain untested. The point is simple. Unless the predictions deduced from a generalization pertain to *actual* and *particular* events or things, they are not evidence of predictive accuracy. That point bears especially on the practice of economists to "test" or "run" a model by using assumed values of the model's constituent variables. The issue is not whether the assumed values are realistic (see Schlefer 2012), nor even when "a model" justifiably translates as "a theory." Rather, whatever their use, *assumed* values cannot be equated with *actual* events or things, economic behavior or anything else. Bottom line: if a purported test of a model is based solely on assumed values, the outcome is not a test of the model's predictive accuracy.

Again, a test of a generalization requires at least one deduced prediction about a *particular* set of events or things. Consider an illustrative case. Among countries, the greater a country's degree of urbanization at any point in any year, the greater is its suicide rate for that year. There are real prospects for tests of the generalization, real because there are conventional formulas for computing the degree of urbanization as well as the annual suicide rate; and requisite data are readily available in published form. Yes, there are various issues: (1) the generalization does not assert causation, (2) there are widespread doubts about the reliability of published suicide rates, (3) the degree of urbanization depends on what is taken as urban units (cities, urban areas, or metropolitan areas), and (4) the degree of association may be far from maximum. But these considerations are strictly irrelevant for present purposes, and the same is true for the negligible theoretical significance of the generalization.

The illustration would be incomplete without considering a quite different generalization. For any country, an increase in *anomie* will cause an increase in the country's annual suicide rate. Sociologists are likely to regard the generalization as plausible, especially if they accept Emile Durkheim's suicide theory (1951[1897]); but they are not prone to recognize that the generalization is far from testable. The point is not just negligible consensus as to causal evidence; additionally, and more to the point, "amount of anomie" defies an agreed upon comprehensible definition, let alone one that promises appreciable empirical applicability.

Back to the core issue: testability. Despite all of the foregoing, the importance of testability will be rejected or questioned in some circles, perhaps driven by what has become widespread hostility or indifference to Karl Popper's thesis (1965) that a theory is not scientific unless falsifiable. Whereas the indifference is common in sociology, Popper receives much more attention in the philosophy of science; and there hostile reactions have escalated in recent decades (Sokal and Bricmont 1998).

The present treatment of testability departs from Popper in two major respects. First, granted that "falsifiability" and "testability" can be construed as equivalent, an untestable theory is nonetheless a scientific theory if it comprises some synthetic statements in the form of logically related generalizations, be they axioms, postulates, propositions, or theorems. Second, it is one thing to characterize a theory falsifiable in principle (i.e., some synthetic statements in the form of generalizations), but quite another to declare that a theory has been falsified. Popper's criterion of a "falsified theory" is here rejected in light of his statement that evidently applies to any theory (Popper 1965: 86). "We shall take it as falsified only if we discover a *reproducible effect* which refutes the theory." The objection is simple: requiring *consistently* negative evidence for falsification of a sociological generalization or theory is scarcely more realistic than requiring consistently positive evidence to sustain the generalization or theory. Moreover, if the predictive accuracy of a generalization or theory relative to contenders is the basis, along with other dimensions of predictive power, for judging its merits, there is no need for the notion of "falsified." Rather, in light of the *relative* predictive power criterion a generalization or theory may be judged (albeit tentatively) as superior or inferior to contenders, perhaps extremely so.

We come then to the basic question: is testability necessary for science? It would be strange to find anyone, let alone scientists and even philosophers, who would deny the necessity, but some of them *might as well*. Especially in sociology, an appreciable number of practitioners promote this adage: the question is more important than the answer. The adage is commonly invoked to protest "dustbowl empiricism" or (horror of horrors) "quantification," but surely there is some connection between "acceptable answer" and "defensible test." If, indeed, the question is more important than the answer, then sociology is little more than a heap of unanswered questions, with practitioners squabbling over which are the most important.

Predictive Accuracy

It is difficult to imagine anyone writing at length on prediction in science without mentioning variables and values (in the numerical sense). However, clarification is needed for several reasons, one being the tendency to think of "a variable" as *exclusively* quantitative, meaning expressible only numerically or conventionally thought of that way. Thus, consider the conventional definition of "degree of urbanization" as the percent of the total population of some territorial unit (e.g., a country or a province) who reside in some type of urban unit—a city, an urban area, or a metropolitan area. The definition justifies treating that concept as a quantitative variable; and no one should be confused when someone speaks of, say, 47.9 as being a value of that variable for some particular country. However, some distinctions are qualitative, and that point warrants emphasis if only because two scientific generalizations may pertain to entities that are quite different qualitatively, examples being planets vs. species and countries vs. individuals. For that matter, generalizations may pertain to what are *qualitative* characteristics of entities, meaning attributes that cannot be described numerically, such as kinds of religion or kinds of kinship systems or even the three types of urban units (*supra*).

The most obvious solution is to use the words "properties" and "attributes" when the characteristics are qualitative, thereby reserving "variables" and "values" for quantitative characteristics. That proposal is all the more defensible because the proposed "qualitative terminology" can be clarified considerably by a simple illustration. "Color" is a property of bodies and "red" is an attribute of that property. The only problem is the infrequent usage exceptions, as when physicists speak of "mass" as a *property* of bodies, though it is expressed quantitatively.

Even accepting the proposed "variable vs. property distinction," the notion of predictive accuracy is a very complex subject. Consequently, if only because of space limitations, subsequent treatment of the notion is predominantly limited to its bearing on sociological generalizations. However, the treatment commences with a point that extends to all scientific fields, though its import can be fully appreciated only in light of the conceptualization of scientific theories in Chapters 6 and 7. All too briefly, the notion of predictive accuracy is limited to generalizations in which all of the constituent words/terms are considered as being empirically applicable (such generalizations are identified in Chapter 6 as "theorems" or "propositions"). Accordingly, because theories are testable, if at all, only through tests of their component generalizations, a theory's predictive accuracy is entirely a matter of what those tests reveal.

The Notion of Prediction

Needless to say, the notion of predictive accuracy presumes another, the notion of prediction itself. As the word is used throughout this book and warrants repeating many times, prediction does not mean foretelling, prophecy, projection of trends, or a guess/hunch about the future. Rather, a prediction is a synthetic statement (the "synthetic" being henceforth understood) that (1) pertains to some characteristic—a property/attribute or variable/value—of an entity or entities as of some particular time point or period and (2) was *deduced* from a generalization or a combination of generalizations about such entities.

One crucial point is that a prediction may be about the future (e.g., the approximate year that the residential population of the U.S. will cease increasing) *or* it may be about the past (e.g., the first year that the crude birth rate of the U.S. declined). As for the "present," it is implied by an assertion about the *existence* of a thing or an on-going event; but they surely commenced in the past, and tests of scientific assertions obviously cannot be based on data not yet collected.

If a prediction is about something in the past, whether it was widely accepted or otherwise "known" in some sense at the time of its deduction is *not* relevant. Nonetheless, there is a crucial question about what is commonly referred to as a "factual observation" (actually an assertion): what generalization or combinations of generalizations, if any, implied the observation and how was it deduced? If implied, the observation is evidence of the generalization's predictive accuracy or inaccuracy. If not deduced, the observation has no scientific relevance, though it may serve as evidence in some future assessment of a generalization's predictive accuracy.

Lest you be puzzled by the denial of the past/future relevance, a major conventional question about any new theory is something like this: what known uniformities does it explain? If deduction is at least necessary for "explanation," deductions about the past play a crucial role in science. Indeed, once it is recognized that literally there is no such thing as "present data," let alone "future data," any test of a generalization or theory necessarily pertains to the past, though the outcome of the contemplated test will not be known until the requisite data have been gathered and analyzed. Perhaps most important, a test of any statement deduced from a generalization is a test also of that generalization, and such is the case even if the test is thought of as "retroactive." The point is not likely to be fully appreciated without recognition that a statement deduced from a generalization may contradict that generalization. Consider this generalization: For any country over time, an increase in the degree of urbanization during a period will be *accompanied by* a decrease in the birth rate. Anyone with knowledge of U.S. trends in urbanization and fertility will recognize immediately that changes in the two variables between 1940 and 1950 flatly contradict the generalization that *implies* the generalization.

The Focus on Generalizations

A full understanding of predictive accuracy requires identification of three components of generalizations and also distinctions as to types of generalizations. The first component of a *qualitative* generalization denotes some kind of entity, the second identifies one or more properties/attributes, and the third is a relational stipulation. An illustration: among human populations identified as hunter/gatherers, there is no institutional slavery. The "Among human populations identified as hunter/gatherers" is the *unit term*, designating the relevant kind of entities in testing the assertion. "Institutional slavery" is a *property* of the entities, with the implied property attributes being "present" or "absent." Finally, "there is no" is the relational stipulation, one that asserts an attribute of all such entities—institutional slavery is absent in all such entities.

Assume that someone examines the anthropological literature and the HRAF with a view to computing the percentage of hunter/gatherer populations for which there was no indication of the presence of institutional slavery. Whatever the reported percentage, it is indicative of the generalization's testability *and* its predictive accuracy. However, various cautionary notes would be warranted. The reported percentage would not be evidence that *other* investigators will report even approximately the same percentage, not even for the same literature. To the extent there is disagreement, the empirical applicability of the words/terms in the slavery generalization becomes questionable.

Types of Generalizations

The kind of prediction that can be deduced from a generalization depends on the type of generalization. The hunter/gatherer instance (*supra*) is a "first-order" generalization, meaning only one property term, "institutional slavery." Although one can think of "institutional slavery" as a matter of extent, the word is used in the generalization as designating a property, meaning it is absent or present as an institution in a population of hunter/gatherers and in that sense a characteristic that cannot be described numerically. Rather, "institutional slavery" in the generalization denotes something simply present for what anthropologists call a "band" and for that reason, the generalization is identified as "qualitative." However, a first-order generalization may be "quantitative" in that the characteristic in question can be described only in numerical terms. Thus, puzzling and questionable though it may be, the following assertion would be an instance of quantitative first-order generalization: regardless of location or historical period, in any city more than 50.0 percent of the marital arrangements are monogamous.

Just as a first-order generalization may be either qualitative or quantitative, so is that the case for all other types of generalizations subsequently identified. However, with a view to reducing complexity, bear in mind that henceforth only the *quantitative* types are examined.

A second-order quantitative generalization asserts an association between two variables, as in: "Among countries, the greater a country's degree of urbanization at any point during a year, the less the country's crude birth rate for that year." Like all of the other illustrations, the generalization was not deduced from a theory; consequently, the rationale for it is obscure, but that point is irrelevant for present purposes.

A third-order quantitative generalization is more complicated. It asserts a relation between some variable and what is a bivariate relation. The form is: among two or more *sets* of entities, the greater the magnitude and/or amount of variance in *W* for a set, the greater (or less) the *intraset* *X*-*Y* relation for that set. Despite the appearance of complexity, the idea of a third-order relation is fairly simple. It is a way by which a theorist asserts that the degree and the direction of an association between two variables (here *X*/*Y*) is contingent on a third variable, its intraset magnitude and/or amount of variance; and the entities in the sets may have been selected so as to maximize or minimize *interset* differences as regards the *intraset* magnitude of or amount of variance in the third variable (here *W*).

One complexity is that a theorist may assert that the *W*/*X* association or the *W*/*Y* association (intraset association in either case) is contingent on the intraset magnitude and/or amount of variance in *W*. Space limitations preclude further explication of a third-order relation; so we will restrict further discussion to second-order quantitative generalizations.

Distinctions Pertaining to Predictive Accuracy

The immediate question in contemplating "predictive accuracy": what are you attempting to predict? There are three possible answers. In the case of an association between variables: first, the association's direction; second, its form; and, third, its degree.

There are only two possibilities as regards the *direction* of the association between two variables, designated here as *X* and *Y*, in a particular test of a related generalization. If the generalization asserts something like "the greater *X*, the greater *Y*" or "*X* varies directly with *Y*," it has *directional* predictive accuracy in that particular test if and only if the *X*/*Y* association measure is positive. Accordingly, if the generalization asserts something like "the greater *X*, the less *Y*" or "*X* varies inversely with *Y*," it has *directional* predictive accuracy in that particular test if and only if the *X*/*Y* association measure is negative.

Virtually any statistical text offers a variety of associational measures that have positive or negative signs (e.g., *r, rho, tau*), and the generalizer should choose among them. However, in giving procedural instructions for a test the generalizer may stipulate that additional variables enter into it, in which case he/she will identify the appropriate multivariate measure of association (e.g., regression coefficients, partial correlation coefficients), the only requirement being that the *X*/*Y* association must be shown as either positive or negative.

Even if the *direction* of the *X*/*Y* association is as predicted in a particular test, the association's *form* may or may not be as predicted; and any generalization should be formulated such as to indicate the expected form. To wit: a positive or negative association may be linear, exponential, or logistic. Again, statistics texts offer procedures for identifying the associational form in particular tests.

As for the *degree* of association, statistics texts also offer a variety of possible measures (partial correlation or multiple correlation in addition to *r, rho, tau*). In formulating a generalization, the generalizer may choose whatever association measure he/she considers as most appropriate, though recognizing the need to satisfy the assumptions in using particular measures (e.g., the association is linear) as stipulated in statistics texts. However, it is also essential to recognize that all alternatives only suggest an answer to this question: for the set of entities in question (e.g., U.S. cities as of 2010), given the value of *X* or *Y*, with what accuracy can the value of the other variable be predicted? Again, conventional associational measures offer no more than a suggestive answer. Thus, whatever the *X* and *Y* values, an *r* coefficient for their association of, say, .150 suggests negligible "precisional accuracy" when it comes to accuracy in predicting an *X* or *Y* value for a particular entity (e.g., the fertility ratio for some U.S. city as of some date in 2010).

Precise predictions, accurate or not, may require an application of a regression equation to the values. A possible exception is a prediction of the rank of some entity (i.e., its ordinal value) from its rank with regard to some other variable; but in the vast majority of cases the ranks have been derived from ratio or interval values, in which case a close association between any rank and the magnitude of the corresponding ratio/interval value (that transformed into a rank) is not even assumed. Moreover, ignoring ordinal values (i.e., ranks), in testing some quantitative generalization only rarely is anyone interested in the predicted value for a particular entity. So precisional accuracy is largely a matter of the *average* error for the set of entities in question. But that average is only one facet of predictive accuracy, and one distinct from the *degree of association* as expressed by some conventional measure (e.g., *r*).

As a brief digression, the authors recognize that the notion of precisional accuracy is truly relevant only for the advanced sciences, physics among others. Moreover, anticipating the charge of "aping physics," we nurse doubts about emphasizing even *degree* of association when assessing the predictive accuracy of sociological theories. Let us put it this way: if a sociological theory generates consistently correct predictions though *only* as to the *direction* of association, it should be an occasion for wild celebration.

On returning to the main subject, although the generalizer is free to choose whatever associational measure he/she thinks most appropriate for the variables at hand, the choice should reflect recognition of gross measurement errors for sociological variables. It can be argued that the impact of measurement error on an association is reduced when the values are transformed from ratio or interval into ordinal values (i.e., rank orders). Of the alternative measures of ordinal association *tau* is especially strategic because it can be interpreted as bearing on proportional reduction in prediction errors (Costner 1965; Frankfort-Nachmias and Leon-Guerrero 2014), and that interpretation is especially advantageous when comparing the predictive accuracy of contending generalizations.

A Special Case

A particularly difficult complexity in assessing a generalization's predictive accuracy (always in a particular test) is the possibility of a partial or fully *nonmonotonic* association, wherein the direction of the *X*/*Y* association changes from negative to positive or vice versa as the values of a variable (*X* and/or *Y*) increase. The subject is important if only because everything said up to this point about predictive accuracy is limited to monotonic associations.

Nonmonotonic association is treated as special case, and the reason is not just that conventional statistics offer no simple method to determine the extent that a bivariate association qualifies, especially whether it is partially or fully nonmonotonic. An association is fully nonmonotonic if and only if the association's direction changes at or near the midpoint value of one of the variables and remains changed from that point onward. In any case, to the extent that an association is nonmonotonic, it should be reckoned as satisfying both the directional criterion and the form criterion. However, the degree or amount of association is another matter, and statistic texts provide instructions for expressing the degree of a nonmonotonic association.

A Final Complexity

Associational accuracy is far more complex than the foregoing suggests. Respect for the reader's patience and awareness of space limitations preclude further treatment of the subject beyond one more complexity.

It is entirely possible that the strength of an association changes over the range of values measured in ways not previously described. For example, universities frequently use SAT scores to predict success in college. In the second quintile and below as well as upper score levels (e.g. fifth quintile) of SAT scores good predictability is achieved, especially with quantitative measures and students who plan to major in engineering. The predictive accuracy of mid-range of scores, however, is not much better than chance. The authors know of no conventional name for such a pattern.

Given the complexity of social life and the stochastic nature of such things as factors related to success in high school, it should be obvious that uncertainty is endemic in the social sciences. Yet exact confirmation is rarely found or expected, and the subject introduces the disturbing suggestion that science should be abandoned in the highly stochastic social realm. However, if so, we should also abandon science in the no-less-stochastic realms of species evolution and epigenetics.

Another disturbing suggestion is that "predictive accuracy" is simply too complex to be a basis for judging the merits of contending sociological theories, even granting that it is really a matter of contending generalizations. Suppose that each of *three* contending generalizations about some variable identifies a *different* independent variable as the basis for predicting the values of the dependent variable. Critics of the "predictive accuracy criterion" could advance two questions. First, what if competitive tests indicate greater *directional* accuracy for one contender, greater associational accuracy for another, and greater *precisional* accuracy for the third? To complicate things further, what if such invidious comparisons hold in one test series but not others? The questions become even more challenging in recognition of alternative ways of detecting or measuring the direction, form, and degree of association.

One response might be to throw up your hands, proclaiming: "This is too much. Let's abandon the enterprise." Another might be to declare, arbitrarily, that one particular facet of predictive accuracy must be paramount in, say, more than two-thirds of the tests. Such arbitrariness would be both obvious and troubling but far from necessary. Granted the complexity entailed in comparing contending generalizations or theories, the goal is a contender that has greater predictive accuracy regardless of the kind considered. As for the likely reply that such a contender will never be found, it would be clearly conjectural and all the more given the distinct possibility that science will never end.

Another question concerning the predictive accuracy criterion: for any given measure of association, what *minimum* value would verify the generalization? The question is pernicious on two counts. First, it ignores Popper's contention (1965), following Hume, that no generalization or theory about an infinite category can ever be verified, certainly not conclusively. Second, if a generalization/theory is judged by its *relative* predictive power, including relative predictive accuracy, there is no need to stipulate any absolute minimum level for any predictive power dimension or any facet of any dimension.

Range and Predictive Power

Scientists study a marvelous variety of events or things, such as planets, riots, earthquakes, volcanoes, families, atoms, suicide, and gravity. That variety points to the inevitability of limits on any scientific theory. So forget about a "theory of everything" and understand that for any particular generalization/theory, the words, terms, or phrases that identify the relevant entities are *unit terms*, and they determine the *range* of a generalization/theory.

As for the connection with predictive power, consider two contending theories about human fertility rates. Suppose that one implies predictions only about international variation ("among countries" as the unit term), while the other implies predictions about *both* international variation and variation among social classes *within* any country.[11] As such, the *range* dimension of predictive power for the second theory would be twice that of the first.

The initial complexity is that a theory may have enormous range despite having only one *explicit* unit term. Thus, one marvel of Newton's gravitational theory is that a "body" may be any *observable* physical entity—a chair, a leaf, an animal, a rock, etc. The range of Einstein's theory is even greater, though for extremely varied phenomena the Newton/Einstein contrast is inconsequential.

Another complexity attaches to the recognition that a single unit term may identify various kinds of entities in the way of subclasses, as opposed to multiple unit terms. Thus, "human populations" designates diverse kinds of entities, such as the residents of two or more cities regardless of location vs. age groups in a particular country. Such diverse possibilities pose a problem when it comes to empirical applicability. Without agreement as to the *subtypes* denoted by the theory's unit term, all reported tests are suspect.

Consider the publications of Talcott Parsons, especially *The Social System* (1951). The works of this once dominant figure among sociology's theorists provide an especially relevant illustration of the empirical applicability problem. "Social system" appears to be the unit term in Parsons' theory; but he never defined that entity in empirically applicable terms, and there are all manner of lingering questions. Are all countries instances of a social system? Does the term apply to all American citizens who identify themselves as "white"? What about "imagined communities" of any type (B. Anderson 2006)? The list goes on. Surely Parsons did not believe that his statements about social systems extend to all such populations; yet the criteria for applying the term remain opaque. Doubts grow on recognition that in Parsons' *The Social System* there are frequent generalizations about "social action."

The point: Sociologists commonly differ as to defensible tests of a particular theory because the theory's range is debatable. Doubts would be minimized by a mode of formal theory construction (forthcoming in Chapter 6), meaning a set of rules for stating a theory, rules that transcend the conventions of a natural language (e.g., German) and call for, inter alia, the stipulation of unit terms. True, theorists in the advanced sciences rarely employ a formal mode of theory construction *other than* the language of mathematics. But those sciences are blessed with terminological conventions and traditions that are alien to sociology, largely because the advanced sciences (including biology) went through a long "natural history stage" and did not borrow terminology uncritically from any natural language nearly as much as did sociology.

Scope and Predictive Power[12]

In testing a generalization values of independent variables are taken as given and predictions are made about the values of dependent variables.[13] The number of dependent variables determines a theory's "scope." To illustrate, suppose some theory implies predictions about international variation in the amount of stratification *and* the amount of political stability, while another also implies predictions about international variation but *only* as regards political stability. Despite identical ranges, one theory's scope is twice that of the other and potentially implies twice the number of predictions, and they would be contending theories.

The importance of scope is not limited to its bearing on "explanatory power." It also addresses a frequently encountered practice of using multiple indicators or purported measure of a single concept. Consider this question: if a theory purports to explain variation in human fertility, what fertility measure should be used to test the theory? The crude birth rate and the fertility ratio are two of several possibilities.[14] However, the theorist need not stipulate one fertility measure to the exclusion of others. Rather, he/she may assert that the theory applies to all types of fertility measures, thereby furthering the theory's scope but risking reduction of *predictive accuracy* (one of many dangers in "doing science").

Defining alternative *versions* of a variable may be a strategic step toward realizing "testability." For example, suppose a theory pertains to crime rates, defined as the reported number of incidences per number of population members. There are three versions of the rate for any of several types of crime, with distinctions pertaining to incidence data—officially reported, victim-reported, or self-reported. However, a theorist need not stipulate all possible versions of a variable. Indeed, whatever the variable, the theorist might exclude those versions that he/she believes would reduce the theory's relative predictive accuracy. So the difficult but attractive goal when constructing theories is to maximize scope without sacrificing predictive accuracy. Is it like a game? Yes!

Does the scope dimension encourage introduction of "absurd" measurement strategies and/or variables? If truly absurd, such introductions would reduce the theory's testability and/or predictive accuracy. By contrast, some version of a variable may maximize the predictive accuracy of so many theories that it becomes conventional, an enormous advantage for any science but all too rare in sociology because of insufficient attention to tests and predictive accuracy.

A Brief Aside

The foregoing brings us to a brief aside on Kuhn's (1996 [1962]) widely read and influential discussion of shifts in scientific paradigms, shifts that he suggests make theories incommensurable. In making that argument Kuhn ignores the crucial point that commensurable theories must be *contending* theories. At least in sociology contending theories have the same unit term (expressed or implied) and share at least one dependent variable, though perhaps different definitions and/or measures of that "shared" variable. Otherwise, sociological theories are *not* contenders and, hence, they are also not commensurable. And such distinctions are by no means peculiar to sociology. Think of Darwin's theory and Einstein's theory; they are *incommensurable* because they are *not* contenders.

As many critics of Kuhn have noted, he complicates matters by strongly suggesting that commensurable theories are an illusion. The problem is not just that Kuhn stopped short of an unambiguous declaration that all theories are incommensurable. Clearly, they are not. With that said, we quickly note a distinction not really examined by Kuhn at all: commensurability is possible and relevant *only* in the case of *contending theories*. [15] It takes only a little searching to find that the history of science is a study in contending theories.

The rationale for this brief aside on Kuhn is simple. He was perhaps the most influential philosopher/historian of science in the twentieth century, and in embracing Kuhn sociologists have been prone to ignore his ambiguities, his varied meaning of "paradigm" (Masterman 1970), and his critics in physics (especially Cromer 1993) along with those in the history/philosophy of science. For example, when addressing Kuhn's arguments David Hull (1988: 492) wrote: "The trouble with the … arguments is that occurrences that should be impossible happen all of the time." As a case in point, the literature of science and philosophy is awash with (1) comparisons of the theories of Newton and Einstein and (2) statements clearly indicating that Einstein introduced a new "paradigm."

Time-space Specificity and Predictive Power

All scientists are at least dimly aware of types of "space-time" relations between variables and that science is little more than the analysis of one type or another. Ignoring substantive considerations entirely, one major type is commonly designated as "cross-sectional," meaning that the relation between two variables, *X* and *Y*, is examined by comparing *two or more* entities (e.g., cities, species), with no concern for the time frame, though most frequently the values of the variables are for the *same* time point or period. For example, using 2010 data we could test a "deterrence generalization": excluding states that have no capital punishment, states with greater estimated certainty of execution (CE) for murder have lower homicide rates (HRs). Alternatively, we could average the *annual* values of both CE and HR for each state over, say, 2009–2011; but the comparison would still be cross-sectional, and the same even if there is a time lag (e.g., 2010 for each CE value but 2015 for each HR or "lagged" average annual values, possibly 2014–2016).

The counterpart is "longitudinal" or "temporal," meaning some relation over time for what is taken as the *same* entity. Thus, we might take California as the entity and examine its CE values and HRs each year over, say, 1995–2015. As in the cross-sectional case, there could be a lag in the years. Whatever, the comparison would be longitudinal because it is the same entity or what is treated as the same entity *over time*.

Why introduce the distinction? Because at least in sociology, research findings clearly indicate that the cross-sectional association may be appreciably greater or appreciably less than the longitudinal association, and even the *direction* of the two associations may differ (i.e., one positive but the other negative). The crucial point is that tests of any generalization may well indicate that its predictive accuracy is extremely contingent on the type of space-time relation asserted by the generalization.

An Elaborate Typology

Hold on to your hats, reader; there is more. Table 4.1 shows the strategic types of relations, both temporal (heretofore "longitudinal") and the spatial (heretofore "cross-sectional"). Thus, in the case of a temporal relation (one entity over time), if the *X* and *Y* values are for the same point/period in time and *neither* expresses amount of change, it is an instance of Type IA1 or Type IA2 in Table 4.1. Otherwise, if the values do express amount of change, the relation is Type IB1 or Type IB2. As for corresponding spatial relations (comparisons of two or more entities), it could be any of the four types in the bottom row of Table 4.1.

Should it appear that Table 4.1 overly complicates, it actually simplifies in that space-time specificity may entail even more distinctions in the way of subtypes. Obviously, the typology would become much more complicated if extended beyond bivariate relations. Less obvious, in the case of Type IB2 or Type IIB2, whatever the temporal metric (nanoseconds, years, or centuries), the change periods *may or may not* overlap. Moreover, an assertion of a "lagged relation" is ambiguous unless it stipulates the lag's length. Should the length that maximizes predictive accuracy be unknown, exploratory research prior to publication of the generalization is needed.

Unless a specific type of temporal relation is stipulated in stating a generalization, it is necessarily and unfortunately left to the discretion of those conducting tests and invites disputes. Consider a generalization that asserts a relation between two variables, *X* and *Y*. Whatever *X* and *Y* may be and whatever the generalization's unit term, the direction and degree of association depends on the space-time character of the association. And note again that the distinctions exceed those in Table 4.1.

TABLE 4.1 Eight Types of Space-Time Relations between Two Quantitative Variables

	No Value of Either Variable, X or Y, Expresses Amount of Change in the Variable		*Each Value of Both Variables, X and Y, Expresses Amount of Change in the Variable*	
Nature of the Entities Considered in Computing Some Measure of Association between the Variables*	For computational purposes the relevant pairs of *X* and *Y* values are such that the *X* value and the *Y* value in each pair pertain to the same point or period of time	For computational purposes the relevant pairs of *X* and *Y* values are such that the *X* value and the *Y* value in each pair do not pertain to the same point or period of time	For computational purposes the relevant pairs of *X* and *Y* values are such that the *X* value and the *Y* value in each pair pertain to the same point or period of time	For computational purposes the relevant pairs of *X* and *Y* values are such that the *X* value and the *Y* value in each pair do not pertain to the same point or period of time of time
One entity that is treated as being the same for all points or periods of time	Type IA1	Type IA2	Type IB1	Type IB2
Two or more entities that are treated as being different	Type IIA1	Type IIA2	Type IIB1	Type IIB2

Source: authors

*An entity could be an individual, a territorial unit (e.g., a city or country), a situation, a condition, a kind of behavior, or any aggregate of the foregoing; but each instance must be thought of as having a spatial limit or boundary.

The principal qualification: in stating a generalization the stipulation of some particular type of space-time relation is important but does not insure predictive accuracy. To illustrate, contemplate two *contending* generalizations about the relation between the same two variables and the same kind of entity (the kind designated by the unit term). Suppose one generalization asserts a Type IIA1 relation between *X* and *Y*, while the other asserts a Type IIA2 relation with each *Y* value one year later than the corresponding *X* value. Tests of the two generalizations could indicate a sharp difference in predictive accuracy. Even so, both generalizations could be "wrong," meaning negligible predictive accuracy for both.

An Implication of Time-Space Specificity

Just below the surface when specifying the "space-time" relation asserted by a generalization, usually there is the issue of causal evidence. True, scientists differ as to what kind of "space-time" relation justifies a causal inference; but that evidentiary issue is beyond resolution, and it need not be confronted even beyond what is said about it in subsequent chapters. We simply note this: in stating a generalization, the theorist should stipulate the type of "space-time" relation that he/she believes is most likely to manifest the presumed causal relation (if any is presumed) and, hence closer than that for any other type. That belief can be assessed by comparing test findings for the specified type of space-time relation with those for any *other* type, even though for the same variables and comparing the same entities (e.g., European countries 2011) or at least the same kind of entities. To treat causation otherwise is to perpetuate a sterile debate and the hoary war with David Hume.

Lest there be any misunderstanding, the manifestation of *presumed* causation is a matter of both space and time. The presumed manifestation is determined by the choices of the unit term and the temporal terminology in the generalization. Those choices establish the range of the generalization, and manifestations of causation are contingent on the *kind* of entities designated by the unit term (e.g., the effects of population density in the case of prisons or schools could be quite different than those for cities, which in turn could be quite different from those in the case of countries). Indeed, correlates of density may well be more contingent on the kind of entities than on the kind of temporal relation if only because appreciable short-run changes in density levels are fairly rare.

Detailed discussion of all time and space contingencies would take us too far afield, but it strains credulity to accept the idea that *all* causal relations, whatever the variables and entities, are manifested in one and only one particular type of space-time relation. Credulity is strained because there is every reason to believe, especially considering the types of social entities (e.g. tribes, families, cities, countries) and varying time frames (moments, weeks, years, decades) considered by sociologists, that the space-time manifestation of a causal relation is extremely contingent on kinds of space-time relations as well as the variables in question. This is one of many *specific* reasons why empirical relations in sociological research could be much more complicated than those in the physical sciences.

Discriminatory Power and Predictive Power

Suppose that a theory comprises two generalizations, one asserting a relation between variables *V* and *W*, and the other asserting a relation between variables *X* and *Y* (the asserted relations may or may not be of the same space-time type). Suppose further that the theory includes this assertion: the *V-W* association exceeds the *X-Y* association. That assertion gives the theory *discriminatory power*, in this case the maximum degree if the predicted difference holds in all tests. So discriminatory power is a matter of amount, and that amount is the proportion of tests in which the *difference* for a pair is as predicted. To illustrate, if a theory comprises four theorems, there are six pairs of theorems and differential association is predicted for each pair. Now suppose that each theorem has been tested three times, in which case there would have been eighteen predictions. Finally, suppose that of the eighteen predictions only eight proved to be correct. At that point the theory's discriminatory power would be .444 (i.e., [8/18]).

What is the basis of discriminatory power? It could be a principle that possibly applies to all theories, such as: the greater the variance in the values of the variables in a particular test situation, the greater the degree of their association. Or it could be something peculiar to a particular theory, wherein the theorist doubts some variables more than others as regards measurement and/or data quality. The last possibility is especially relevant for sociology. Only rarely can a sociologist realistically assume anything like maximum reliability for any measure or data, and even rarer will a thoughtful theorist assume even more or less the same reliability level. The more general point is that maximum discriminatory power is the mark of a sophisticated or very lucky theorist.

Parsimony and Predictive Power

Although conventional, the notion of a parsimonious theory is difficult to define, especially on recognition that (1) even if equated with "simplicity," parsimony is a matter of degree and (2) agreement in independent comparisons of contending theories will be rare. Yet parsimony is frequently mentioned as a major goal, as evidenced in an interview with Steven Weinberg (NOVA 2003: 1) who noted with his usual conciseness:

> The whole aim of fundamental physics is to see more and more of the world's phenomena in terms of fewer and fewer and simpler and simpler principles. And the way you do this is not by having one book on electromagnetism, and another book on the weak interactions, and so on, but to have just one book on all the forces of nature. A simpler description–that's what we're aiming at.

Whether parsimony is sought or not, we offer this definition: a theory is parsimonious to the extent that its conclusions (theorems) exceed its premises (axioms, postulates, and/or propositions). Some connection between parsimony and predictive power is undeniable on recognition that a theory's premises imply predictions through theorems in the form of hypotheses. Accordingly, the greater the ratio of theorems to premises, the "more from less" principle holds.

Our definition of parsimony presumes that a theory's premises and theorems can be identified with confidence. Unfortunately, those components of a typical sociological theory cannot be identified readily. Indeed, exegeses of theories flourish in sociology primarily because the logical structure of the typical sociological theory is obscure.

The obscurity will prevail as long as theorists employ *only* the conventions of a natural language, such as English or German. Those conventions promote vagueness and frequently result in idiosyncratic deductions. The more immediate point is that parsimony can be judged with greater confidence if the theory has been stated in accordance with rules that transcend the conventions of any natural language. Such rules are likely to require, *inter alia*, explicit identification of the theory's components (e.g., axioms, postulates, propositions, and theorems).

Granted some possible relation between parsimony and elegance, the empirical applicability of elegance as a property of theories is extremely doubtful. Some scientists are enchanted by the notion of elegance, as one might stand in aesthetic appreciation of a painting. It is important, however, to recognize the subjectivity involved. In praising some theory for its elegance we may or may not be assuming its testability and impressive predictive accuracy, both luxuries that sociologists rarely if ever enjoy.[16] Finally, an elegantly coherent theory is not necessarily parsimonious, and the desirability of coherence is not lost in adopting relative predictive power as the criterion, not even the exclusive criterion, for assessing theories. After all, incoherence diminishes testability, and it likely reduces predictive accuracy even when it does not preclude tests.

The Larger Question

Some inquisitive reader is likely to pose this question: why are you fellows so obviously preoccupied with the notion of predictive power? We would be sorely tempted to reply that it is definitely not a whim, but a serious and informative answer is required. The "informative" reflects the need for an answer that goes beyond predictive power to some other notions in our conception of science.

Answering the question as briefly as possible, an appeal to predictive power is the most effective way to reduce uncertainty about empirical assertions. To be sure, there are other possible ways to reduce uncertainty, prayer being one of them for the religiously inclined. But no other institution rivals science when it comes to focusing on empirical assertions, and uncertainty is more of a constructive force in science than in any other institution.

Surreptitous Criticism

Well, there you have them—the seven dimensions of predictive power. Opponents might pose a follow-up question: what is the order of importance? The question could be surreptitious criticism, an anticipation of some arbitrary ordering. That anticipation is understandable, but the dimensions need not be ordered.

Whatever the subject—gravitation, kinship systems, genetic properties, homicide rates, etc.—the ultimate goal is a theory that exceeds contenders with regard to *all* dimensions of predictive power. As for the contention that the goal will never be realized, it assumes that science will someday end. To the contrary, at least as defined here, science is likely to endure as long as humanity survives.

Even if eventually there are accepted theories on all subjects, someone will ask: do the theories differ as to predictive accuracy and, if so, why? Still another perennial question: if a generalization or theory appears to have impressive predictive accuracy despite maximum range, how so? Such questions rightly suggest that as long as humans survive, there is no reason for science to end (Horgan [1996] notwithstanding); and the same holds for discoveries (Stannard [2010] notwithstanding).

Final Considerations

The paramount feature of the proposed definition of science is *prescriptive*: A scientific answer to a question *should be* assessed in terms of the answer's predictive power. Lest that feature be misunderstood, five considerations warrant special treatment, though, their order is again inconsequential.

Five Considerations

First, should our definition of science and its defense be dismissed as "positivism," the response would by necessity be obscure. Over thirty years ago Peter Halfpenny (1982) demonstrated uses of the label "positivism" that amount to twelve distinct meanings, which alone justifies abandoning the word. Pejorative use is like a dog barking in the night; hostility is manifest but the target is obscure.

The second consideration is more complicated. For over fifty years, recent critics have subjected science to numerous critical appraisals. Perhaps the best known are Thomas Kuhn (1996 [1962]), a physicist cum historian of science, and Paul Feyerabend (1975), a philosopher of science.[17] In addition, as previously suggested, Rupert Sheldrake (2012: 6) argues that "science is being held back by centuries-old assumptions that have hardened into dogmas." Two of the ten dogmas Sheldrake addresses (2012: 7) will suffice as illustrations. "Everything is essentially mechanical … The laws of nature are fixed." He is actually seeking an expansion of the range and scope of theories to subjects (e.g., telepathy) that he considers alien to the "materialism" of contemporary science. Would the "expanded theories" be assessed in terms of their predictive power? His misgivings about contemporary science fall far short of an answer.

The third consideration is an emphatic disclaimer. An immediate or conclusive assessment of a generalization or theory in terms of predictive power is rarely if ever feasible. It is not just that neither testability nor predictive accuracy should be judged on the basis of one test. In any scientific field, outcomes may vary substantially among test situations, and errors and even fraud are always possibilities. Obviously, that variation will not be revealed in one test, and there is no warrant for assuming that test findings will be even approximately uniform over time. Even when test outcomes are at least somewhat uniform over time, disagreements as to the predictive power of some theory (especially relative to contenders) are far from impossible. So why tout predictive power when it does not eliminate doubt in assessments of theories or generalizations? Simply this: it promises more consensus than does any alternative. That contention is especially supported by the case of sociology, where diverse criteria have flourished for centuries in the form of assessments fueled by opposing perspectives—functionalism, Marxism, feminism, etc. Those perspectives conspicuously share only one thing, indifference if not hostility to making relative predictive power the ultimate criterion for assessing theories, and some writers (Schwandt 1996) reject the very idea of criteria.

The fourth consideration is best expressed as this claim: The predictive power criterion is far less radical than it may appear. Such notions as evidence, facts, and correspondence with reality are staples in both science and philosophy, and they are clearly related to testability and predictive accuracy. Moreover, those two dimensions are conventional concerns apart from making predictive power the *supreme* criterion. As cases in point, testability is evidently the foremost concern in negative assessments of "string theory" (see, especially, Woit 2006), and no one denies that the predictive accuracy of quantum theory promoted its acceptance.

The fifth and final consideration relates to traditional criteria for judging theories. When someone praises a theory for its explanatory power, surely he/she cannot reject the proposed range/scope criteria. Nor does it distort to equate simplicity and parsimony. Finally, as for specificity and discriminatory power, only space limitations preclude a demonstration that those notions are consistent with conventional critiques of scientific theories.

A Vast Lacuna

By far, the most neglected subject in this chapter is suggested by Richard Feynman's statement, quoted in Chapter 1 and here repeated. "The principle of science, the definition, almost, is the following: *The test of all knowledge is experiment.* Experiment is the *sole judge* of scientific truth."

Again, astronomy and geology especially create doubt about Feynman's declaration. Additionally, it would be highly unlikely and even more ethically inappropriate to carry out experiments on such things as the deterrent effects of capital punishment. Nevertheless, non-experimental evidence can be brought to bear on the deterrence doctrine, and all too little has been done in the way of field experiments.

Despite the foregoing reservations, the importance of experimentation cannot be gainsaid; so a defense of virtually ignoring the subject is imperative. Because a thorough treatment of *experimental science* (the design of experiments and assessments of their possible outcomes) would require a book in itself, the following is limited to the connection of the subject with predictive power.

In one sense or another, experimentation may be a larger component of science than are theories, and experimentation may well be the only basis for testing *some* theories. Thus, it is difficult to imagine any purported test of a version of "string theory" that will not be experimental (see Gorham 2009: 46–52, and Woit 2006); indeed, the foremost criticism is its resistance to experimental tests.

It would be difficult to exaggerate the role of "curiosity experimentation" in scientific discoveries. While it may be the most underrated kind of scientific activity, the goal of the other major kind of experiment—one undertaken to refute or corroborate something—is inseparable from prediction. Indeed, perhaps the signal scientific virtue of well-designed experiments is the control they offer over *contending* predictions. Granted uncertainties in even the most carefully designed experiments, repetition of them is science's trump card; and the very idea of repetition entails comparative tests. To be sure, observational studies can be repeated, but experimental repetition is of greater evidentiary importance because it offers more opportunities for controlling conditions (e.g., manipulating "range" and "scope") and the connection with prediction is obvious. Bottom line: whether observational studies or experimentation, prediction is the central evidentiary criterion for science.

Notes

1 In the social sciences, see for example: Alice Goffman 2014, especially Appendix, A Methodological Note; Phillip E. Hammond (ed.) 1967; Hortense Powdermaker 1966.

2 For early critiques of the "externalist view of science," see Chalmers 1990 and Hull 1988.

3 The reader may want to peruse the HRAF website for current information and an indication of how the files have been used. www.hraf.yale.edu.

4 Reduced to its essentials, for any field a scientific perspective should: (1) delimit the field's subject matter, (2) formulate the kinds of questions that are to be asked, and (3) stipulate the criterion or criteria for acceptable answers. This list is virtually identical with Keith Parsons's identification (2014: 32) of the key components of Kuhn's "paradigm," an identification less ambiguous than Kuhn's conceptualization.

5 Contemplate Richard Miller's comment (1987: 15): "The dispute over the nature of explanation is one of the most heated in the philosophy of science."

6 Nevertheless, many writers on science make reference to causation and/or use causal terms without recognizing problems. One recent case is Paul Thagard's *The Cognitive Science of Science* (2012). Much more is said on "causation" in Chapter 8.

7 Contemplate the platitudinous objection of David Deutsch (2011: 14), a physicist: "...prediction is not, and cannot be, the purpose of science." Of course, it is not *the* purpose of science. Whatever that purpose, predictive power is the supreme criterion for assessing attempts to realize it.

8 The title of Karl Popper's *The Logic of Discovery* (1965) is grossly misleading, in that his "falsifiability" criterion actually amounts to a logic of justification. Indeed, much of the criticism of Popper may stem from the yearning of his critics for a logic of discovery. A genuine instance would be

marvelous, but the creator would be foolish to publicize it. Rather, he/she should use that logic to crank out theories and reap endless awards.

9 There are at least two possibilities for a logical relation between theories. First, they pertain to the same kind of entity and/or share at least one property/variable. Second, the kind of entity and/or the property/variable in one theory is a subclass of a more inclusive one in another theory (e.g., city residents/human population and robbery rate/total crime rate in another).

10 A preoccupation with anticipating the future may stifle science. Consider Wade's observation (2014: 218) on China in the 1600s: "The Chinese had little curiosity about astronomy itself. Rather, they were interested in divination, in forecasting propitious days for certain events, and astronomy was merely a means to this end."

11 Illustrating another way, the keen desire of physicists for a testable theory that combines Einstein's theory and quantum mechanics is a yearning for a theory with greater range. Then consider Weinberg's observation (2009: 13) on thermodynamics. "The laws apply to black holes, they apply to steam boilers, and to many other systems. But they don't apply everywhere. Thermodynamics would have no meaning if applied to a single atom."

12 Both scope and range bear on Lee Smolin's contention (2006: xiii): "In science, for a theory to be believed it must make a new prediction–different from those made by previous theories..." That contention is consistent with the present emphasis on relative predictive power; and even though Smolin does not clarify "new prediction," it surely has something to do with the range and/or scope of a "new theory" (whatever the criterion of novelty).

13 A causal language is not essential for the independent/dependent distinction.

14 Crude birth rate is generally taken to be the number of live births occurring during some calendar year, per 1,000 population estimated at midyear. The fertility ratio is taken to be the number of children 0–4 years old compared to the number of women in child bearing years, conventionally between 15 and 44 years old.

15 Consider two questions related to Weinberg's (2009: 192) interpretations of Kuhn: "Kuhn argued...that the paradigms that govern successive periods of normal science are *incommensurable*." Granted incommensurable paradigms, are *contending* theories possible only within the same paradigm? For that matter, how can there be contending theories within the same paradigm? Answers would be disputable, and the reason is that Kuhn's treatment of theories and paradigms is a study in ambiguities. This latter contention is fully consistent with Keith Parsons's fair-minded critique of Kuhn, but Parsons (2014: 32) muddied the water with this statement: "Kuhn calls this core theory or set of theories the 'paradigm'..." A paradigm is or can be a theory?

16 Ian Glynn's *Elegance in Science* (2010) demonstrates that *illustrations* of purported "elegance" are a far cry from an empirically applicable definition.

17 Negative views of conventional science thrive all the more in the humanities and literary world (see, especially, Curtis White [2013]).

5

CONCEPTUALIZATION OF A SCIENTIFIC THEORY

What we have at this point is a definition of science (Chapter 4) that implicitly focuses on what scientists do. When doing science, scientists gather information and formulate assertions about the characteristics of entities. These assertions are the building blocks of science, possibly culminating in "theories," perhaps the most sought-after product.

Theories are constructed along many paths. In a frequently quoted comment, written for the occasion of Max Planck's 60th birthday, Albert Einstein (Einstein 1918) concluded: "There is no logical path to these laws; only intuition, resting on sympathetic understanding of experience, can reach them." Undoubtedly, Einstein would have said the same about paths to a scientific theory, and our definition of a scientific theory (*infra*) is consistent with his perspective. No less important, the definition applies to all of the sciences, including sociology.

Major Conceptual Problems

So what is a scientific theory? Five interlinked problems confront any attempt to answer. First, in many instances the question is greeted by silence, as scientists who promulgate theories are not prone to define them. Second, the word "theory" is used in several fields, philosophy and mathematics in particular, wherein the practitioners do not identify themselves as scientists. Third, upon close examination, various candidates for a scientific theory in the social/behavioral sciences are a mixture of hunches, speculations, and interesting insights, with few if any testable implications (consider, e.g., the putative theories of Sigmund Freud, Karl Marx, Talcott Parsons, and Max Weber). Fourth, definitions commonly offer little by way of separating the meaning of theory, a theory, scientific theory, and *a* scientific theory.

The fifth problem, flowing from the first four, is the prevalence of broad definitions of "theory," those that simply equate a theory with an idea. This prevalence is aptly described by Sjoberg and Nett (1968: 29), a description that holds after nearly fifty years:

> One group of social scientists would identify, at least implicitly, any kind of conceptualization with theory. Such concepts as "status," "role," "culture," or "public opinion," when defined and utilized in the interpretation of research materials, are equated with social

> theory. In this sense, theory comes to be identified as almost any thought process as opposed to the observation of facts per se.

Note particularly that the Sjoberg/Nett passage pertains to "theory," as opposed to "a theory" or "theories." It would be comforting to think that sociologists differ far less when defining "theory" than "a theory" or "theories." The implied comparison would be difficult if only because of the paucity of *theories* in sociology's literature over recent decades. Definitions of "theory"—partial or full, implied or explicit—are much more common, and their diversity cannot be exaggerated (see Tindall's critique [2000] of items in the sociological literature). Some indication of that diversity is provided by Abend's description of *seven* "different senses of the word 'theory' in the sociological language." (2008: 176). Unfortunately, having promised (2008: 173) to "tackle" the question of "what ought 'theory' to mean in the sociological language," Abend eventually demurs. Why he does so seems clear. Having advocated "ontological and epistemological pluralism" throughout, why would Abend violate it with a definition? He wrote fully in the spirit of Paul Feyerabend's *Against Method* (1975), and his failure to describe specific "political mechanisms" (2008: 173, 194) that supposedly would "help" end this particular chapter of conceptual anarchy in sociology is understandable. A thorough treatment would smack of orthodoxy, an obvious threat to pluralism.

Eight Definitions

Given the apparent disinclination of scientists and even philosophers of science to define "*a* theory" and the loose specifications when they do, we again turn to dictionaries and encyclopedias to seek alternatives. The eleven publications examined while seeking definitions of science (Chapter 4) became the source of definitions of "theory." The rationale for examining the same publications was the possibility of some relation between definitions of "science" and definitions of "theory." No particular relation was detected.

Of the eleven publications wherein there is a definition of science, in three instances (Hagstrom 2000; Walker 1999; and Shermer 2005) there is no entry or subentry for "theory."[1] The remaining eight are in Appendix II, but the reader need carefully examine only the three quoted below, as they were selected to represent distinctive types. Each type is described in terms that indicate why any instance is judged as defective.

Definition I

At the risk of appearing uncharitable, we identify Barnhart's (1988: 1132) definition as an instance of the "extremely ambiguous" type. It reads: "The sense of the principles or methods of a science or art rather than its practice is first recorded in Robert Cawdrey's *A Table Alphabeticall* (1613). The sense of an explanation based on observation and reasoning is first recorded in 1638."

The phrase "science or art" clearly indicates that the definition applies to an indescribable variety of kinds of human activities even if they are limited to what is commonly identified as "fields." Surely mathematics and philosophy qualify, but the practitioners do not identify themselves as scientists or artists, even though they may refer to the products of their activities as "theory" or "theories." As to the content of "theory," Barnhart's definition is misleading at best. In particular, whatever one means by "method," why is it a component of a field's "theory"? Equally relevant, although any field can be thought of as having "principles," that word scarcely applies to *all* of the axioms, postulates, propositions, or theorems of what is commonly regarded as *a* scientific theory.

Finally, the second sentence in Barnhart's definition can be construed as inconsistent with the first in that it appears to be a definition of *a* scientific theory, not of "theory." Whatever, by any conventional criteria the definition appears grossly incomplete. Even if the definition were extended, no amount of elaboration would placate those critics who, along with Einstein quoted above, would object to the suggestion that scientific theories are necessarily the product of observation and reasoning, a suggestion of some logic of discovery that excludes "pure intuition" as a possible route to a theory.

Definition II

We label the second type of definition, again perhaps uncharitably, as the "hodgepodge" type. An example comes from Pickett (2011: 1805):

> 1: A set of statements or principles devised to explain a group of facts or phenomena, especially one that has been repeatedly tested or is widely accepted and can be used to make predictions about natural phenomena. 2: The branch of a science or art consisting of its explanatory statements, accepted principles, and methods of analysis, as opposed to practice: *a fine musician who had never studied theory*. 3. A set of theorems that constitute a systematic view of a branch of mathematics. 4. Abstract reasoning; speculation: *a decision based on experience rather than theory*. 5. A belief or principle that guides action or assists comprehension or judgments: *staked out the house on the theory that criminals usually return to the scene of the crime*. 6. An assumption based on limited information or knowledge; a conjecture.

If you are puzzled or confused by the definition, you are not alone. It is as though the author defined at least three different things, though far from completely. Specifically, the second sentence evidently relates to "theory"; the third and fifth more nearly pertain to "*a* theory"; and the remaining sentences are partial demarcations of "*a* scientific theory."

Observe that "scientific theory" is apparently left undefined. That omission could reflect the belief that the very notion is indefensible; but "scientific theory" is a meaningless term unless it denotes *nothing more* than all of the theories of a particular scientific field, if not all sciences. Even so, the word could be useful for some purposes, and the same may be said for "theory" (versus "*a* theory").

Definition III

We now come to the third type, which we construe as "ambivalent." Ellis and Swoyer (2008: 343) provide the example:

> The notion of a theory is controversial in social science. A single and simple conception of theory is unlikely to apply across all fields, from mathematical economics to cultural anthropology. Still, construing *theory* broadly as any attempt to systematize and explain certain phenomena, it is clear that theories play a central role in social science.

This definition is narrow, being focused on the social sciences. Given the title of the publication in which the definition appears, the limited character is understandable. Yet there is no indication as to why a "single and simple conception of theory" is unlikely to apply to all fields, though perhaps an unfortunate possibility. Indeed, would the restriction of a definition to each particular field eliminate controversy? If so, how and why?

What appears to be the major merit of the definition—the focus on "*a* theory" in the first sentence—becomes doubtful by the switch to "theory" elsewhere. The switch is all the more puzzling because it is difficult to see how "any attempt to systematize and explain certain phenomena" can be "theory" rather than "*a* theory." For that matter, the use of the word "systematize" is more likely to puzzle than clarify, all the more in suggesting an activity rather than a product.

Commentary on All Eight Definitions

Many of the criticisms of these instances of three types of definitions extend to some if not all of the others in Appendix II. If you are like the authors of this book, you are likely to regard the definitions on the whole as difficult to understand, perhaps even more than the three just considered. The difficulty stems in part from confounding four distinct subjects—theory, *a* theory, scientific theory, and *a* scientific theory. Remembering that we relied on dictionaries and encyclopedias, it appears that some conceptualizations pertain to all four of the subjects. The authors of the definitions in Appendix II may understand such terminological distinctions, but only two seem to recognize them explicitly.[2] Specifically, in *Definition* 7 the parenthetical "(without the article)" could be an elliptical recognition of "a theory" vs. "theory"; in *Definition 8* there is "a theory" in the first sentence, only to be followed by "theory" in the second sentence; and in the last sentence we find "theory" and "theories."

More on Definitional Issues and Problems

The foregoing stops short of implying that issues and problems can be resolved or circumvented by avoiding a definition that applies indiscriminately to all four subjects—theory, *a* theory, scientific theory, and *a* scientific theory. To the contrary, there are many seemingly insoluble problems and irresolvable issues in defining any one of the four. Nevertheless, if only to simplify, from this point on, unless indicated otherwise the exclusive concern is with "a scientific theory" or "scientific theories."

Theories versus Isolated Statements

Whatever else "a scientific theory" may be, it comprises more than one statement.[3] The justification for the claim is that the typical putative theory in science comprises (1) at least one synthetic statement (be it designated as an axiom, postulate, theorem, proposition, generalization, principle, or hypothesis) and (2) at least one analytic statement that defines some word or term in the synthetic statements, perhaps a procedural prescription or a formula.

The Relational Question

Given the idea that a theory necessarily comprises multiple statements, what is their relation? There is no obvious alternative to something like this all too brief rendition: a theory comprises statements that are logically related if only through common words, terms, phrases, and/or definitions. Thus, if at least one such component appears in each of two or more statements, there is a logical relation among the statements.

The foregoing approaches an important issue, especially in sociology. As argued in Chapter 6, the vast majority of theories in sociology are stated in accordance only with the conventions of a natural language (e.g., Spanish), and those conventions are far too limited and ambiguous to establish indisputable logical relations within a particular set of statements other than, possibly, "common components." As to what constitutes a "logical relation" *beyond* the common-components criterion, that subject is treated in Chapter 6.

The Analytic/Synthetic Distinction

Repeating an earlier characterization, an analytic statement is true or false by the very meaning of its constituent words and/or its syntax.[4] The illustrative instance is taken from Steven Pinker (2009: 13): "A bachelor … is simply an adult human male who has never married." Virtually everyone would agree that Pinker's statement is *not* an empirical assertion, for it is true by definition. The point has nothing to do with the possibility of alternative definitions (e.g., deleting "adult"), nor with the empirical applicability of Pinker's definition.[5] Rather, contra Quine, some statements can be identified confidently and correctly as "analytic"; indeed, any definition is analytic. However, lest confusion ensue, while definitions are analytic, some analytic statements are not definitions, an instance being: if a population does not cease growing, its size will double. Given the conventional meaning of the word "doubling," the statement could not be false. Reversing the issue, also note that analytic statements may be *false* by the very meaning of constituent words or syntax. Again, Pinker suggests an example: Some married men are bachelors. The statement is false and by definition.

Synthetic Statements

At this point, the reader might understandably ask: why beat the subject to death? Enough already! We proceed with some regret and even apology, but as necessary if only because in jousting with "empiricism" Quine (1951, 1960) evidently denied the analytic/synthetic distinction. Critics of empiricism or positivism frequently and gleefully invoke that ostensible denial, and it bears on definitions of a theory.

Rejection of Quine would be incomplete without reconsidering the nature of a synthetic statement. Again repeating, a synthetic statement is a statement that is *not* true or false by the very meaning of its constituent words and/or its syntax. Consider an illustrative instance: Average age at death is less for married men than for widowed men. How could anyone demonstrate that the statement is true or false by the conventional meaning of "widowed men" and "married men" or any other component of the statement? Certainly, tests of the statement might prove difficult and the outcomes disputable. At most, tests would corroborate the statement only in some populations, and the reliability of official mortality statistics by marital status is questionable; but that litany of problems has no bearing whatever on identifying the statement as "synthetic."

Despite what has been said about the analytic/synthetic distinction, critics might still invoke Quine's seeming denial. The reaction (Chapter 4) to that denial warrants repetition. First, if Quine's argument means that the distinction cannot be confidently applied to *any* statement, then a multitude of scientists have wasted much of their life attempting to test some generalization or theory. Second, if his argument means that the distinction is not somehow indisputable in *some* instances, only the incorrigibly naïve would dissent.

An important related issue can be stated as a question: must a scientific theory necessarily include definitions? A negative answer is suggested by some definitions in Appendix II; but even an explicit answer (negative or affirmative) would not resolve the issue, though one consideration points to a negative answer. The natural history stage of some science could be so long and constructive that the words, terms, or phrases in some if not all eventual theories need not be defined (i.e., the conventional meaning insures sufficient empirical applicability). However, that possibility should not be confused with the use of a "primitive term" in a theory's premises, in which case the meaning of the term (1) is implicitly defined by its relation with other terms in the premises, (2) defies a clear definition, or (3) is assumed to be empirically applicable without being defined.

The "Testability" Issue Again

When a statement is identified as synthetic, the identification suggests that the statement is testable in principle, directly or indirectly. Various issues and problems pertaining to "testability" were examined in previous chapters, particularly Chapters 3 and 4, and in recognition of their importance they are treated again in Chapter 6. Here it will suffice to point out that some of the definitions in Appendix II do speak to the issue.[6] In particular, the first component of *Definition 3* requires not just testable statements as constituents of a theory but also some support for the theory, though several of the other definitions do not speak to the question. Whatever the case, in the few instances where scientists have defined a scientific theory, they tend to explicitly or implicitly *exclude* widely "rejected theories." Consider Dennis Trumble's definition of a scientific theory (2013: 46): "an explanation for some aspect of the natural world that is well supported by observed facts and tested hypotheses."

In connection with defining "*a* scientific theory" there are various issues pertaining to testability and test outcomes, including "testability not relevant." But why would anyone deny that testability is necessarily relevant? The question applies especially to "string theory" in physics (Smolin 2006) but also to the tradition of "grand theory" in sociology (e.g., those of Marx and Parsons). The testability of grand theories is disputable at best, but rightly or wrongly they are commonly identified as "scientific."

It is refreshing to point out that the testability issue haunts the physical sciences as well as sociology. Nevertheless, why are reports of systematic tests of sociological theories so extremely rare? There are several reasons, the most immediate being the difficulty of distinguishing analytic and synthetic statements in sociological theories, a veritable "Quine feast." That issue is examined at greater length in Chapter 6.

References to Explanation and/or Causation

We come now to perhaps the stickiest issue of all: must the definition of a theory include words relating to explanation and/or causation? Even the "and/or" is disputable, especially if it is granted that an explanation of some thing or event does not necessarily identify a necessary *and* sufficient cause.

Problems and Issues Pertaining to Explanation

All definitions in Appendix II support the contention that scientific theories are explanations. The word explain, explanatory, or explanations appears in each of them. However, an inspection of widely accepted theories in science will reveal that not one of the three words—explain, explanatory, or explanation—appears in them. Such lacuna is not surprising in light of this seldom expressed contention: we *read* explanations into theories and generalizations.

Those who read explanations into theories or explicitly require it of theories should be obliged to again confront Miller's contention (1987: 15): "The dispute over the nature of explanation is one of the most heated in the philosophy of science." That claim can be extended to science in light of a statement by David Deutsch (2011: 3), a physicist: "Scientific theories are *explanations*: assertions about what is out there and how it behaves." Deutsch (2011: 30) does offer a more elaborate definition of "explanation," but it is scarcely an improvement. Such statements are common in the science literature, and they suggest that scientists should let philosophers wrestle with the nature of explanation. After all, scientists could be content with something like this: our theories assert some particular order in events and things, and do not demand more than that unless you really want to dive into a sea of problems.

The Evidentiary Problem in Connection with Explanation

Consider a generalization in this form: of two phenomena, *X* and *Y*, an instance of *X* always appears before and in the vicinity of an instance of *Y*, while an instance of *Y* never appears without an instance of *X* preceding in that vicinity. That is to say, *X* is both necessary and sufficient for the appearance of *Y*. For one reason or another, many commentators would deny that the generalization implies an adequate explanation of *Y*. Such denials would persist even if the generalization is altered to further specify the temporal and spatial relation of *X* and *Y*. Indeed, rejections of the generalization's explanatory adequacy are likely to persist even if there is abundant evidence that the asserted *X*/*Y* association holds without any known exception.

Note the distinct possibility that the reasons given for rejecting the generalization are secondary to the point that the critics would have read "explanation of *Y*" into the generalization. However, should the reader regard the illustration as unrealistic, we encourage you to read extensively in the abundant literature commencing with Carl Hempel's "covering law" requirements of an explanation (Hempel and Oppenheim 1948). The general point is not just that evidence of explanatory adequacy is debatable; additionally and more important for present purposes, the very notion of explanation gives rise to many issues and problems apart from "explanatory adequacy." Should you, the reader, nurse doubts about the magnitude and/or number of issues and problems, try your hand at formulating a definition of "explanation"; and then go on to specifying criteria for an adequate explanation. Finally, even if you are satisfied with your "product," contemplate the variety of possible objections to it.

And how do sociologists confront the problems and issues pertaining to explanation? Well, they have a marvelous strategy—just ignore them. That strategy is clearly revealed by Douglas Marshall (2008: 212–214). He does say: "At issue is just what it means to 'explain' a phenomenon." But far from resolving the issue, at no point does Marshall even set forth an explicit definition of "explanation," let alone criteria for assessing the adequacy of an explanation. Evidently just say theories "explain" and let it go at that.

Problems and Issues Pertaining to Causation

Establishing a causal connection is even more fraught. The use of a causal language has been treated at various points previously; so only brief mention need be made here. Since David Hume it has become obvious that anything like effective agreement cannot be realized either in defining causation or in formulating empirically applicable criteria for casual evidence. And never mind the hoary war with Hume, including endless symposia. For centuries he has survived his critics.

The question we pose is simple: when defining or formulating "a scientific theory," why incorporate words, terms, or phrases that clearly appear conducive to insoluble problems and irresolvable issues? In particular, given that difficult and controversial questions about causal evidence have lingered on for hundreds of years, why force that notion into a definition? Do not invest a huge amount of time contemplating the question, as there is more to come.

Issues Posed by the Finite/Infinite Distinction

A generalization may be limited to a finite category of events or things. Thus, an assertion about the morale of Confederate troops at Gettysburg would certainly be a generalization, but it would be finite, limited both spatially and temporally. The question becomes: does a scientific theory necessarily comprise at least one synthetic statement about an *infinite* category of events or things? Restating the question in terms of statements: does a statement—explanatory or not, causal or otherwise—warrant scientific attention only if it applies to an infinite category of events or things, unlimited temporally and/or spatially?

Regardless of how expressed, the question is extremely controversial for sociology.[7] All too briefly, numerous sociologists—especially Weberians, ethnomethodologists, phenomenologists, and social interactionists—evidently have grave doubts about any theory pertaining to infinite categories of socio-cultural phenomena (e.g., all revolutions—past, present, and future). Indeed, those doubts may be the principal basis for rejecting physics as the model of science. By contrast Donald Black (2011: xii) is quite explicit in his claims for explaining conflict across time, place, and life circumstances:

> The purpose of the theory of moral time is scientific: to explain conflict. Although the theory is not perfect or final, I believe it is extremely powerful. It clearly and simply explains the vast majority of conflicts I have examined over the years, not only in modern societies but across the world and throughout history. It pertains to conflict in literally every kind of human relationship, whether between relatives, friends, colleagues, strangers, groups, or nations.

Observe that Black favors theories in the form of generalizations about infinite categories, which in the terminology of *predictive power* translates as "maximum range" (Chapter 4). The emphasis points to a crucial difference between Black and the authors of these pages. Although Black occasionally wrote favorably of "testability" and frequently used the word "predicts," he has not clearly advocated *relative predictive power* as the sole criterion for assessing sociological theories, much less explicitly identified the relevant dimensions. Had he done so, his crusade to drive reductionism, individualism, and teleology from sociology would not have been necessary (even if feasible). If such "things" are theoretically sterile, as Black consistently implies, then their use in formulating theories is bound to result in less predictive power than that of some other theories, those that expressly or tacitly reject such things. However, in that connection, Black's numerous critics have scarcely offered contending theories, much less any that remotely rival the ostensible predictive power of Black's. So we admonish his critics: put up or shut up.

Particular Features of the Issue

Although the word "general" appears in several of the eight definitions in Appendix II, there is nothing like an explicit statement pertaining to the finite/infinite distinction. This lacuna may simply reflect the belief that all scientific theories have an infinite quality, but that belief is hardly convincing if sociology is taken as a science.

As understood here, the finite/infinite distinction is actually more nearly a matter of degree, with the "range" of a theory's assertions being the most relevant consideration. Recall from Chapter 4 that "range" is a dimension of predictive power; as such, it is seen as a contingent property of theories, not a definitional component. In that regard it is interesting to consider whether the range of Darwinian theory reaches beyond Earth and, if so, how far? Likewise, the conclusions of cosmologists may someday create doubts about assertions of an unlimited range in the generalizations of physicists, and "black holes" alone pose a range question about some of the generalizations in physics. So, even ignoring all of the social sciences, there are justifications for *not* making infinite range a necessary feature when defining a scientific theory.

An Alternative Definition of a Scientific Theory

So where do the foregoing ruminations lead? Rather than resolving issues or solving problems, our proposed definition *circumvents* them. Specifically, in our definition we avoid the use of three words—explanation, cause, and infinite—or related terminology. *A scientific theory is a set of logically related assertions about contingent characteristics of some designated entity or entities, with the understanding that the assertions are to be assessed by their predictive power relative to that of contenders.* Clearly this definition calls for numerous clarifications if only to further its empirical applicability.

Contingent Characteristics

The notion of "logically related" has been clarified previously. It will be reconsidered in the next chapter in more detail, but the idea that a scientific theory comprises "assertions about contingent characteristics of some designated entity or entities" warrants immediate attention. The first clarification is purely terminological. The word "characteristic" is used as a generic designation. It translates as any property/variable or the related attributes/values of an entity or entities, and it is contingent in not being identified by the definition of the entity ("not true by definition" being the conventional expression). Although an assertion may pertain to only one characteristic of the entity (including "existence" or "existential" as a possibility), it more commonly pertains to some kind of relation between characteristics. And note especially that, unless indicated otherwise, the reference is always to *contingent* characteristics.

An entity is any event or thing or category of them. So the estimated diameter of Mars is a quantitative characteristic of an entity, a particular planet, and the number of planets in a solar *system* is a quantitative characteristic of an entity (the solar system in question). By contrast, "presidential assassinations" is a qualitative characteristic of entities *as events*—that being assassinations—while Lincoln's assassination is a qualitative characteristic of an entity in the way of a particular event. But there are entities in the form of physical *events*, such as a particular "oceanic earthquake" (a qualitative characteristic—"oceanic"—of a phenomenon); or the *number* of detected earthquakes in California during the last decade (a quantitative characteristic of a particular entity—California). Finally, "official homicide rates" may appear to be a category of things, but one could argue that it is really a category of events, with "official homicide" being a qualitative characteristic of rates.

The last example suggests a complexity heretofore ignored. Stated all too briefly, agreement in the use of the foregoing terminology, characteristic and entities in particular, is likely only when considering statements, especially in the form of assertions. Stated more simply, it is a matter of context. Consider the assertion "Official homicide rates are more reliable than official suicide rates." In the context of that assertion, both the first three words and the last three words are entities, with "more reliable" being an asserted *characteristic*. But consider another assertion in the form of a generalization: "The official homicide rates of Mississippi are greater than those of Vermont." In that context "official homicide rates" are characteristics, here of two states.

Whatever the ambiguities of the terminology in our definition of a scientific theory, they are largely due to the extensive applicability of the definition; it is intended to apply to all of the sciences. As for sociology, consider the following assertion in the way of a generalization: among countries, the greater the degree of urbanization for a country, the greater the degree of social stratification in that country. In light of the proposed terminology, "countries" are entities and both "degrees" are quantitative characteristics. No sociologist is likely to be confused by those designations, nor fail to recognize that for present purposes the testability and predictive accuracy of the generalization are irrelevant.

In light of the foregoing, any assertion about a characteristic or characteristics of any entity (or entities as a category) is an actual or potential subject of scientific inquiry. However, variation among *entities* (e.g., species, cities) as regards some characteristic or over time for some entity (be it a class or category) is the foremost concern in any scientific study, whatever the science. Thus, Gregor Mendel (1822–1884) studied generational variation in properties of plants (color in particular), and had publication of his finding received the attention it deserved at that time (scarcely noticed for something like fifty years) the science of genetics might well have developed much more quickly. Then for centuries physicists have analyzed variation in the mass of objects. As for psychology, the very notion of personality presupposes variation; indeed, without variation, there would be no personalities, much less theories on the subject. As for the social sciences, despite occasional and sterile interests in such putative constants as "human nature" (i.e., an implicit denial in variation), it does not distort to describe the practitioners as virtually preoccupied with variation. Cultural differences preoccupy many anthropologists, and rates of various kinds are fodder for sociologists. The specific kinds of variation that interest economists and political scientists are less obvious, but they are surely interested in one kind or another.

The basic claim conveyed in our definition is this: if a theory that does not clearly pertain to contingent characteristics, it is not a scientific theory. As for the widely accepted belief that a scientific theory answers questions, it does not follow that the questions are expressly stated in the theory itself (i.e., in one of the premises or conclusions). After all, generalizations in the form of axioms, postulates, propositions, and theorems are the essential components of a theory, but they are not questions. The same may be said of definitions as *possible* components of a theory.

The point is not just that we humans read *both* questions and answers into theories; additionally, such interpretations can be contentious. The theorist may seek to avoid misinterpretations by expressly identifying the questions that he/she thinks the theory answers, but that effort may fail. For that matter, critics may regard the questions as being misleading if not spurious, in which case the answers are moot.

The Centrality of Contingent Characteristics

Any well-known theory in the social/behavioral sciences is commonly thought of as making assertions either about causes of something or its consequences. The other possibility—assertions about *both* causes and consequences—is seldom recognized, though Marx's and Freud's theories are evidence that the dualism is more than just a logical possibility. Much of Freud's theory pertains to alleged causes and consequences of childhood sexuality, the Oedipus complex in particular, while Marx made all manner of claims about the causes and consequences of "economic relations." The real possibility of such a theory is the foremost rationale for the using the word "characteristics" in our definition of a scientific theory, but it should be quickly added that *asserted* characteristics, whatever the entity or entities, must be *contingent* characteristics.

As for the word "characteristic," it is the generic designation of the major words, terms, or phrases in definitions of properties or variables (Chapter 4). However, a contingent characteristic *can be* (1) an asserted cause of something, (2) an asserted consequence of some thing, or (3) some type of space-time relation. Described still another way, a contingent characteristic is commonly thought of as a correlate of some event, thing, or category of them. The crucial point: There is no need to use "causes" or "consequences" in defining a scientific theory, and referring to "contingent characteristics" allows us to avoid the Gordian knot when attempting to formulate criteria of causal evidence.

Of course, the exclusion of a causal language will not go down well among scientists who have become captivated by the bromide "correlation is not causation." They appear blissfully unaware of the point that you may have correlation without causation, but you cannot have causation without correlation of one kind or another (i.e., depending on the type of space-time relation in question).

Particulars versus Categories

Is a scientific theory about some particular event or thing (e.g., the American Revolution or Adolph Hitler) possible, or is it necessarily about an infinite category (e.g., all revolutions, all dictators)? The question is especially strategic and controversial for sociology, but the Copernican theory and the theory of plate tectonics question bring into question the "infinite requirement." For that matter, the infinite requirement negates the possibility of a scientific theory about any unique event or thing, one ostensible instance in cosmology being the "Big Bang" theory.

Our definition admits the possibility of a scientific theory about particular events or things, as well as one about a category (type or class), be it finite (a spatial and temporal limit) or infinite (unlimited temporally and/or spatially). A defense of that inclusive character was undertaken in discussing the dimensions of predictive power (Chapter 4), "range" in particular. For now, we simply say that the issue is by no means limited to sociology, though it is especially acute there.

The Reference to Relative Predictive Power

The lengthy treatment of predictive power in Chapter 4 points to a simple question: when scientists promulgate a theory, do they assume their assertions will be assessed in terms of predictive power relative to that of contenders? They do if they have any hope of eliminating contending theories in light of some objective criterion.

Even if a theory has no known rivals, surely the theory's creator(s) envisioned eventual invidious comparisons that relate to predictive power. It is even more incredible that any scientist is totally indifferent to predictive power, granted the possibly of considerable variation among scientists in that regard. Should it appear that physicists who formulate versions of string theory are totally indifferent, that appearance ignores the dimensional quality of predictive power. Specifically, perhaps "stringers" in physics are willing to sacrifice testability, at least in the short run, for gains in range and/or scope. Stated otherwise, they may be captured by the prospect of eventually having a theory with overwhelming predictive power. It needs hardly be said, they are playing for high stakes.

Avoiding Issues and Problems

To repeat, our definition of a scientific theory does not purport to resolve issues or solve problems. Rather, it was formulated with a view to avoiding them. Along that line, at the risk of beating a dead horse, we return one last time to causation and explanation.

Avoidance of Issues Pertaining to Causation and Explanation

Two brief additional observations will suffice. First, even if causal terms appear in a scientific theory's premises, the conclusions are not testable until the causal terms have been *translated* into assertions of space-time relations. Second, if some type of space-time relation is asserted to be greater than *any other* type (i.e., maximum specificity), the predictive accuracy of that assertion will bear on causation. Moreover, it will not raise an irresolvable issue (the very worst kind) unless critics are (1) determined to continue the war with Hume and (2) balk at granting that space-time manifestations of causation are contingent on the kind of entities in question and/or the properties or variables in question. More will be said on this issue in subsequent chapters.

Another avoidance of an irresolvable issue is the omission of the word "explanation." Yes, surely, many scientists, including sociologists, insist that a theory is an explanation or it is nothing (see Abend 2008: 185, on Homans), but they ignore a previous contention: we *read* explanations into theories.

Ignoring that contention is conducive to ignoring two others, both of which have been treated at length previously. First, it is doubtful that even effective consensus can be realized in defining "explanation." Second, it is even less likely that scientists or philosophers can agree even appreciably in stating evidentiary criteria as to an adequate explanation. Of course, both contentions can be ignored when making reference to "explanation" in stating a scientific theory, but only a taste for obfuscation justifies that practice. Nevertheless, abandoning the practice would not preclude invidious comparisons of scientific theories. To say it again and again, a scientific theory should be judged in terms of its predictive power relative to that of contenders. Even if critics insist on the supremacy of a particular dimension (e.g., testability), it is difficult to imagine them denying that the ultimate goal is a theory that exceeds contenders as regards *all* predictive power dimensions (Chapter 4).

The Analytic/Synthetic Distinction

The third circumvention of an irresolvable issue is perhaps the most distinctive feature of our definition of a scientific theory. We might appear to justify applying the label "scientific theory" even to a set of statements that are analytical (i.e., no synthetic statements). But such a theory would not be testable because no genuine predictions could be derived from it, and it could hardly survive in comparisons with contending theories as to relative predictive power. Specifically, a defensible assessment of a theory's predictive accuracy is possible only to the extent that the theory is testable.

So why tolerate even the possibility of an untestable theory? Well, to be candid, the tolerance reflects a grudging recognition (contra previous opposition to Quine) that the analytic/synthetic distinction can be difficult and controversial. The tolerance reflects another and more constructive recognition. Demanding synthetic statements can be unconstructive if it discourages the kind of conceptual or experimental work that *might* eventually result in tests of a heretofore untestable theory.

The Finite/Infinite Distinction

The fourth circumvention relates to a very contentious question. Stating it again but in another way, does a scientific theory *necessarily* comprise at least one statement about an *infinite* category of events or things? To repeat, the proposed definition points to a negative reply, and it is more radical than it may appear. Even if a scientific theory must comprise at least one synthetic statement, it need not be a generalization about an infinite category. Hence, even though a set of statements about the Spanish Civil War pertains to a finite category of events, the set would *possibly* qualify as a scientific theory, so defined.

The "possibly" is a crucial qualification. A set of statements about a finite or infinite category is not a scientific theory unless made in the understanding that the statements are to be assessed in terms of relative predictive power, for present purposes the *range* dimension in particular. The limited range of statements about a finite category makes them potentially or actually inferior to a theory that comprises generalizations about the same kind of events or things but as instances of an infinite category (e.g., all civil wars).

Final Considerations

Failure to recognize that our definition is limited to "*a scientific theory*" would be an obfuscation in not recognizing that "a theory" may not be scientific. So mathematicians and philosophers need no permission to continue using the word "theory" as they see fit; but it is significant that they do not identify their products as "scientific theories," much less display explicit concern with relative predictive power. What criteria should they employ? That question is their problem!

The Related Contemporary Question

In light of our definition of a scientific theory, a contemporary question warrants attention. Is what passes for "creationist theory" a scientific theory?[8] The question has been debated for decades (see Shermer 2005), and the vast majority of putative "scientists" answer with an emphatic negative; but as far as the authors know that answer is not defended by appealing to an explicit and ostensibly complete definition of a scientific theory that makes relative predictive power the supreme assessment criterion. Nonetheless, in light of our definition at least some versions of creationist theory would qualify as "scientific" but only *if* its defenders grant that any version should be judged solely by its predictive power relative to that of contenders.

Creationists do answer questions about the origins of life, and even some critics might grant that those answers do imply predictions (e.g., as to the Earth's age). However, those critics could rightly make a crucial point: the predictive power of any creationist theory (and not just its predictive accuracy) is negligible relative to any version of Darwin's theory. Trumble (2013: 86) expresses the argument another way. There are various assertions that, *if accepted as valid*, might eliminate natural selection as the indispensable explanatory mechanism in Darwin's theory but support some version of creationist theory. One such possible illustrative assertion: the Earth is less than a million years old.

The point of the foregoing is that Trumble's argument bears on the relative predictive power of Darwinian theory (whatever the version), and that point is more germane than caustic remarks about creationism. Trumble (2013: 86–89) is also especially relevant in replying to the charge that Darwinian theory is far too plastic (i.e., subject to modification) to qualify as "scientific." In that connection, although not suggested by Trumble, it would be wise to speak of a *Darwinian perspective* rather than a theory, meaning a set of generalizations (e.g., all organisms share some kind of genetic material) awaiting integration as a truly deductive scientific theory, though not such as to be considered "complete and final." The point is relevant if only because generalizations, Darwinian or otherwise, should be assessed in terms of their relative predictive power.

The paramount issue is the seeming determination of "Darwinian fundamentalists" to consider some version of Darwin's theory as being long since complete and final, a determination possibly fueled by fear of appearing to support creationism. If so, it is surely a high price to pay for appearance.

A Final Issue

A final issue can be introduced by thinking about a hypothetical set of statements that qualifies as a scientific theory in light of our definition. Even if the statements appear clearly untestable or as clearly inferior to some contender as regards other dimensions of predictive power, the set would remain a theory and no less "scientific."

To be sure, the relative predictive power of some particular theory may appear so negligible that it can be justifiably ignored in surveys of contending theories. Nevertheless, if it has been published, such a theory ought to sit there indefinitely awaiting reassessment. Limited research resources and/or widespread doubts as to plausibility may prevent reassessment; even so, the success or failure of a revival attempt should be determined by judgments in terms of prospective predictive power.

The objective is not eventual absolute consensus in judgments of relative predictive power, whether that of a new theory or an old and perhaps discarded one. Some disagreement is a virtual certainty; it is simply evidence of science's adversarial character. Without doubt, all institutions—religion, government, art, etc.—manifest adversarial qualities, but science differs in that disagreements are much more limited by a focus on dimensions of predictive power. Moreover, science gets along, so to speak, without coercive dispute settlements, in particular the incarcerations, tortures, and executions that haunt the history of some institutions. Indeed, discussions of the difference between religion and science should commence with that particular difference in mind. Unfortunately, that difference is *not* now ancient history, as demonstrated by the current prevalence of sectarian homicide, a prevalence that should sober those American fundamentalists who call for "more religion in government."

Notes

1 Note especially that the dictionary and encyclopedia definitions pertain to "theory" *without* separate entries for "a theory," "scientific theory," or "a scientific theory." As subsequent discussion indicates, an enormous problem is created by the absence of separate entries.

2 The same extends to Jennifer Bothamley's (2002: vii) *misidentification* of an entry in the *Shorter Oxford English Dictionary* as a definition of "*a* theory" when it is really a definition of "theory." Consequently, she recognizes that the definition is a hodgepodge but not the reason for it (i.e., "theory"

does not translate as "a theory," much less "a scientific theory"). That criticism has no bearing on the *omission* of many well-known theories from Bothamley's *Dictionary of Theories*, one egregious instance being Durkheim's suicide theory, the subject of uncountable publications.

3 But contemplate David Papineau's terminology (1979: 24): "... I shall use 'theory' to [identify] any *(lawlike) generalization* or interrelated set of such generalizations." (Italics added.)

4 Here and elsewhere, if the reader understandably balks at the "true/false" distinction, simply substitute "corroborated/refuted." The more general point is that scientists need not enter the jungle of logomachy treasured by philosophers.

5 The point is evidently lost on Pinker (2009: 13), who confounds the empirical applicability of the word "bachelor" in contrived hypothetical situations with the word's empirical applicability *in general*. In that connection, a word or term may be clearly empirically applicable even though left undefined. One kind of such an undefined word or term is conventionally known as "primitives," and their appearance in theories is inevitable.

6 Here and elsewhere, unless indicated otherwise, there is a crucial assumption when a particular Appendix II definition relates to "theory" or "theories." The assumption is that the definition's author was writing about "a scientific theory" or "scientific theories." Needless to say, in several instances the assumption is debatable, and that point underscores the principal shortcoming of the Appendix II definitions in general.

7 Here and many other places, the question could be extended beyond sociology to all social sciences; and an affirmative answer might preclude considering history as a science. Suffice to say, the authors have enough problems without risking the wrath of what could be many historians.

8 Virtually everything said subsequently about either theory, creationist or intelligent design, applies to the other. For one of many devastating critiques of "intelligent design," see Robert Ehrlich (2003: 41–77). By the way, critics of the "design theory" are prone to allow the theory's advocates to get away with a fast one when advocates argue that finding something as complicated as a watch clearly suggests "design." The advocates should be forced to confront this question: what is suggested if we find millions of *broken* watches scattered hither and yon? For that matter, when there is evidence that at least 95 percent of some line of products eventually simply cease "working," what are we to make of the designer? Omnipotent or caring but surely not both.

6

FORMAL THEORY CONSTRUCTION

Illustrations, Problems, and Issues

We come now to something rarely encountered on sociology's landscape. Few sociologists have been attracted to formal theory construction (FTC), some even viewing it with disdain, perhaps perceiving it as a threat to free-wheeling interpretations, if not an attempt to impose orthodoxy. By contrast, those who see merit in FTC claim that it promotes terminological distinctions and testable theories. Elaborating, a mode of FTC offers a systematic vocabulary and a way to link abstract notions with data, both contributing to the principal benefit, greater testability. Scarcely less important, formally stated theories are the most effective way to integrate heretofore unconnected generalizations, usually stated as isolated propositions.

Appearances perhaps to the contrary, the purpose of this chapter is not to promote a frenzy of formal theorizing in sociology. Instead, our argument is that sociology will forever remain at best a grossly incomplete, backwater science unless a substantial number of its practitioners commence gathering data, crafting generalizations, and stating theories in accordance with some formal mode. So what is FTC?

Before we proceed to an answer, a reminder: *a scientific theory is a set of logically related assertions about contingent characteristics of some designated entity or entities, with the understanding that the assertions are to be assessed by their predictive power relative to that of contenders.* Given that definition, a mode of formal theory construction (MFTC) is a set of rules for stating a scientific theory, wherein some or all of the rules transcend the conventions of a natural language (e.g., English, Russian).

Advocating rules that transcend the conventions of natural language encounters resistance, the charge being that such rules open the door to jargon and obscure abstractness, with the ultimate dread being "orthodoxy." Consequently or not, from the field's very beginning, virtually all sociological theories have been stated in accordance with only the conventions of some natural language, wherein distinctive components are the familiarly recognized sentences and paragraphs. Now for a likely reaction: although theories stated in a natural language lack a clear logical structure, stating them in accordance with some formal mode is both difficult and dull. So why bother? Stating the answer as another question: if a natural language is adequate for systematic deductions, why did logicians invent the syllogism and why is mathematics strategic for so much of science? The answer is that both logic and mathematics are based upon clear rules of deduction. By contrast, if a theory is stated in some natural language, there can be no confident identification of premises and conclusions, let alone how the conclusions were deduced.

Natural language conventions also make it difficult to detect inconsistencies in a theory, perhaps blatant contradictions. Moreover, those conventions obscure a crucial connection between deduction and testability.[1] Specifically, it may be possible to deduce testable predictions *systematically* from a formally stated theory even though each premise is clearly untestable in any direct sense. That point bears especially on sociological theories in the grand tradition (e.g., those of Spencer and Pareto). In the rare instances where the premises can be identified confidently, it is risible to suppose that they can be tested systematically in any direct sense.

The formulation of any scientific theory is defective if it fails to explicitly distinguish testable and untestable statements, though all statements are in the form of assertions or definitions. That distinction does not necessarily require FTC, but a formal mode should be such that its use forces the theorist to clearly separate testable and untestable statements. However, identification of a statement as "untestable" creates a problem in the way of a question: how can evidence be brought to bear on untestable statements? The most defensible way is to deduce testable statements from the untestable and do so through a MFTC.

The conventions of a natural language, English or otherwise, are far too limited and ambiguous for systematic deductions. So a MFTC should stipulate rules for deducing testable conclusions (via hypotheses deduced from testable theorems) despite untestable premises (axioms and/or postulates), and mathematics does not offer appropriate rules for all fields, sociology in particular.

Needless to say, the foregoing does not answer a very wide variety of important questions about FTC. Answers are best given by considering a particular formal mode.

An Illustrative Mode of Formal Theory Construction

How common are untestable statements in the sociological literature? It suffices to say they are beyond counting. Consider the term "class conflict" (the "social" is understood). Conflict between classes is asserted in numerous arguments by sociologists, none of whom is likely to deny that the amount of class conflict varies among entities (e.g., countries, cities). Nevertheless, a sociologist would have to be deranged to assume some widely accepted procedure for measuring the *amount* of class conflict, and it would tax credulity to deny that the characteristic is a variable (i.e., a quantitative characteristic). Hence, any generalization about class conflict is immediately suspect as regards testability. This is where some MFTC is really needed, and our proposed mode is set forth in this chapter to demonstrate, sociological tradition to the contrary, that a theory about class conflict can be stated such as to be intelligible and promise appreciable testability.

The Immediate Problem: Defining Conflict and Class

Indifference to conceptual problems, especially empirical applicability, is virtually a sociological skill. But even ignoring empirical applicability, an intelligible definition of class conflict presumes no less intelligible definitions of "conflict" and "class." The authors know of no definition of "conflict" in the sociological literature that promises even appreciable empirical applicability (see, e.g., Collins 2012 and Coser 1956), and an empirically applicable definition of class conflict presumes the same for "class." The sociological literature on class perhaps exceeds that on any other subject, but the many stabs at defining "class" have been defective.

Before examining common defects, note a possible strategy in formulating a theory about class conflict. Given all of the problems in attempting to realize an intelligible and empirically applicable definition of class conflict, one can simply use the term but leave it undefined, a practice that has been frequently pursued by sociologists in writing on the subject, though perhaps unwittingly. Described in the language of FTC, in formulating a theory one can use the term "class conflict" as a *construct* and one left undefined (i.e., a primitive term, though not merely a matter of syntax). But the term should be embedded in a set of premises that imply at least one testable generalization, and positive outcomes in the tests would *indirectly* support the premises about class conflict. We have employed that strategy in offering an illustrative theory on class conflict (*infra*) to explicate our MFTC, and the initial step is a brief treatment of the most glaring problems in defining class.

The Most Glaring Problem

Avowed Marxists in sociology are prone to insist that the very idea of "class membership" is irrelevant in defining a class. Nicos Poulantzas (1975: 17) is the most explicit: "Social classes are not empirical groups of individuals ..." So criteria of class membership simply become irrelevant in defining "class." But if the reality of a class has nothing to do with the identification of its members, what is it? Well, Poulantzas (1975: 170) takes a stab at an answer: "Classes exist only in the class struggle." That declaration is a real howler unless one can demonstrate that classes "struggle" before they exist. To the surprise of no one, Poulantzas does not even attempt that demonstration.

Poulantzas's locutions become more understandable in light of statements of another avowed Marxist. Erik O. Wright (1978: 3–4) describes "class" as sociology's "one independent variable," but without expressing wonderment he recognizes four contending conceptions of class and refers to class as an "essentially confused concept." Later Wright and his colleagues (1982: 709) are even more emphatic: "Anyone familiar with the recent history of Marxist thought will know, there is hardly a consensus among Marxists over any aspect of the concept of class."

Criteria of Class Membership

Most sociologists probably reject the belief that a class is somehow divorced from the characteristics of its members—the superorganic conception. Yet rejection of that conception implies the need for a method to identify class membership.

Three methods (or criteria) have been recognized if not accepted by sociologists for decades; they are known as (1) the objective, (2) the subjective, and (3) the reputational. They can be described readily; but the reader must look elsewhere and early (e.g., Broom, *et al.* 1981) for a treatment of the complexities, and even more extensive treatments commonly ignore various issues and problems, nor are all treated in what follows.

There are three so-called objective criteria for identifying an individual's class—occupation, income level, and educational achievement. The immediate problem with the first criterion is that even in a single metropolitan area there are thousands of specific occupations; consequently, it is imperative to group them into categories identified as "classes," the most common being lower, middle, and upper. The immediate problem is that the grouping of occupations is bound to be a study in arbitrariness, commencing with this difficult question: why those categories-cum-classes rather than others? Less obvious, what criteria are the "categorizers" to use in assigning occupations? If it is income level or educational achievement, the choice between the two will smack of arbitrariness; and the use of either one cannot escape the arbitrary, as suggested by this illustrative question: what is the minimum annual after-tax income or minimum educational level for an "upper-class occupation"? For that matter, if income and/or education are the bases for categorizing occupations, why bother with occupations in assigning individuals to classes?

As for the "subjective" criteria, numerous surveys have demonstrated the feasibility of asking individuals to identify their class, but there are some serious problems. If given a free choice to label themselves, survey respondents give an indescribable variety of answers, such as working class, middle class, above average, and poor. Such responses can be categorized so as to match the terminology of sociologists who write about "class." But is a respondent's identification of herself or himself as "working class" to be translated as "lower class" or as "proletariat"? So it is that the subjective approach creates doubts about the *social* reality of classes and invites arbitrary interpretations of responses to survey questions. One possible solution is to restrict response choices to a list of classes, such as lower, middle, or upper. But that procedure creates anther problem, one illustrated by survey findings pertaining to U.S. class composition. Americans are much more prone to identify themselves as "working class" than "lower class." So limiting respondent choices appreciably determines the class composition of the population surveyed, a defect not remedied by random sampling of respondents.

The "reputational" approach can be readily illustrated. Given a group of, say, only three members, each member can be asked to identify the class of the others. The assumption is that each survey respondent in a particular population knows the other members sufficiently to identify their class. Consequently, whatever the merits of the reputational approach, it is feasible only for very small communities. Moreover, framing the survey questions and classifying the responses given in a "reputational survey" poses all of the difficulties recognized in describing the "subjective" approach (*supra*).

"Old Karl" is not the Solution

The fundamental problem with any of the three methods is that it presumes a fixed number of classes, whatever the label applied to them. Described another way, each method identifies the class membership of individuals but not the identification of the classes themselves or makes the latter identification arbitrary.

Recognition of the fundamental problem could be the reason for insistence of some sociologists (e.g., Nicos Poulantzas) that the existence of a class has nothing to do with the identification of its members. That insistence is in the tradition of Karl Marx (1818–1883), but it is puzzling nonetheless. Avowed Marxists are not prone to recognize that Marx (or an associate) wrote less than two printed pages toward completion of "The Classes," the last chapter in Volume 3 of *Capital* (American publication much later, in 1909). True, Marx's defenders can admit that he left the notion of class very muddled but argue that he correctly identified particular "historical classes," as in this passage (Marx 1909: 1031): "wage laborers, capitalists and landlords form the three great classes of modern society resting upon the capitalist mode of production." What happened to the self-employed (independent farmers and craftsmen in particular) and government employees/officials? Spare us the dodge that Marx was concerned only with the "great classes." Any attempt to identify classes that considers only ownership of the means of production and/or position in economic relations is bound to promise insufficient empirical applicability or bristle with arbitrariness.

Finally, assuming that a class can be identified without considering its members—again the superorganic conception—surely it cannot be analyzed for some scientific purpose without considering the behavior of its members. As a case in point, Erik O. Wright's admirable work (1979) on class structure and income structure was based on *statistics for individuals*. Needless to say, "class structures" do not have incomes, and income without behavior is exceedingly rare.

A Typology of Terms Used in Formal Theory Construction

To clarify the notion of FTC we need to draw some terminological distinctions and later introduce a related typology of *statements* as components of theories. Both the distinctions and the typology are explicated by using them subsequently in stating a theory formally.[2] The theory is not intended to be impressive; rather, the primary purpose is to demonstrate how systematically testable theories can be realized.

Just as the grammar of a natural language is largely based on terminological distinctions (e.g., nouns vs. verbs), so should a MFTC. We begin with a type of term that has no obvious counterpart in the English language.

Unit Terms

Contemplate a statement (clearly an assertion) made in accordance with the proposed MFTC:

Postulate 1: Among countries, the greater the class conflict in a country, the greater the income inequality in that country.

The first *two* words are the "unit term," which identifies the *kind of entities* that would be compared in an ill-advised attempt to test the statement, whether a test is conducted in accordance with some stipulated procedure or in the form of general observations. Unfortunately, the vast majority of empirical assessments in sociology are nothing more than general observations, often on something akin to nebulously defined "human nature."

If a theory is stated in accordance with the proposed MFTC, a unit term appears in all of the theory's premises and conclusions (theorems). If naysayers doubt the necessity of explicit unit terms, they are surely inexperienced when it comes to testing sociological theories or generalizations. The point is not that sociologists refrain from using unit terms. To the contrary, in formulating a theory the typical sociologist uses such a bewildering variety of unit terms that it is difficult to determine what comparisons are to be made in testing the theory. And even when a theorist makes fairly consistent use of a particular unit term, its empirical applicability may be questionable. As a case in point, examine translations of Durkheim's theories and then attempt to answer this question: What would and would not be instances of "society"?

Further clarification of the unit term notion is pursued later in connection with the integration of theories. For the moment greater clarification is realized by returning to Postulate 1. The theorist may have selected "countries" as the unit term because he/she (1) considers the term as sufficiently empirically applicable and/or (2) believes that the asserted association may hold *more* for countries than for other entities (e.g., "cities" or "cities within the same country").[3] In any case, unless one is genuinely confused, questioning a theorist's choice of a particular unit term reflects doubts as to plausibility, perhaps without regard to any actual tests of the theory. Doubts about a theory's unit term may reflect something approaching an aesthetic assessment, such as the demand for intellectually satisfying theories (e.g., Tittle 1996: 347), though such judgments are rarely made explicit. If relative predictive power—as explicated in Chapter 4—is accepted but *only* as one of multiple criteria, why add others (e.g., elegance, intellectual satisfaction) that do not promise anything like effective consensus in assessments of particular theories?[4]

Returning to Postulate 1, "Among countries" is quite different from "For any country over time," and anyone who doubts the evidentiary implications knows all too little about sociological research. Specifically, whatever the variables in a generalization, the spatial relation ("among countries" being an instance) and the temporal relation ("any country over time" being an instance) may be quite different, even possibly the *direction* of association (i.e., "greater … greater" vs. "greater … less").

The implications of different types of space-time relations as regards both casual inferences and testability cannot be exaggerated, though commonly overlooked in the centuries of futile jousts with David Hume over his treatment of the notion of causation. Whereas the subject of causation is a thicket of issues, there can be no denying that the "range" of a theory, a dimension of its predictive power, is determined clearly by the theory's unit term.[5] The point warrants emphasis because sociology's history has been largely one of sterile debates over causation, either in general or particular instances.

Note that Postulate 1 does not even suggest a definition of "a country." Beyond possibly noting that a colony or possession does not qualify, the theorist might leave the term undefined, treating it as a primitive term, meaning that he/she assumes sufficient empirical applicability without a definition. In any case, the theorist may have refrained from "Among societies" as the unit term because sociologists (perhaps taking a lead from Durkheim) have used the term in a bewildering array of meanings.

Finally, should the theorist elect to define "country," the definition would be found in what the theorist *explicitly* identifies as the "extrinsic part of the theory," along with definitions of other terms and possibly procedural instructions (e.g., measurements, accessing data, formulas, or equations). By contrast, the "intrinsic part" of the theory comprises the principal synthetic statements—axioms, postulates, propositions, and theorems. Each type of statement is subsequently defined, and note that one advantage of FTC is that the theorist need not offer such definitions. They come with the package, so to speak.

Although alien to the conventional form of sociological theories, the clear-cut separation of the two principal parts of a theory—intrinsic and extrinsic—promotes testability by explicitly separating the analytic and synthetic statements as components of the theory. Anyone who doubts the need for such a separation is indifferent to testability and/or has never struggled with the prose of sociologists who write in the tradition of "grand theory" (e.g., Max Weber) or one of their exegetical paladins (e.g., Talcott Parsons on Weber).

Characteristics

Those who speak or write in English describe things by using adjectives, such as "red" or "brown," both of which are *attributes* of the property known as "color." As the illustration suggests, property/attributes pertain to *qualitative* characteristics of entities, meaning that they cannot be described numerically or at least conventionally they are not so described. For example, in Postulate 1 "class conflict" and "income inequality" are not adjectives in the strict sense, but they do pertain to quantitative characteristics (here, of countries), "quantitative" in that no one is likely to deny that both are a matter of amount.

Because characteristics are frequently quantitative, one may be tempted to equate "property" and "variable" as types of characteristics. This is understandable, but there is a tendency to think of "variable" as somehow divorced from a "unit term." Yet an utterance in which there are words/terms that designate one or more variables but no unit term would be virtually meaningless. Think of this illustrative utterance in which *X* and *Y* are variables: The greater *X*, the greater *Y*. Whatever *X* and *Y* may be, even if the utterance is taken as an assertion, it borders on the unintelligible. Now suppose that *X* is "residential density" and *Y* is "homicide rate." The reasonable inference would be that both terms are variables and pertain to human populations. But what kind of human population? The possibilities are virtually infinite—residents of a census tract, an urban neighborhood, a city, an urban area, a metropolitan area, a state or province, a country. Furthermore, even someone with no training in criminology might suspect (and rightly) that the association between the two variables depends appreciably on the kinds of entities considered in calculating the association in question, and the "over time" association for any of those kinds of entities (e.g., the city of Boston, each census year, 1910–2010) could differ enormously from the corresponding spatial association (e.g., among the counties of California, values of the two variables as of 2010).

So "unit terms" make generalizations empirically meaningful, possibly testable, and a major determinant of predictive accuracy. This response is anticipated: but natural language conventions make explicit unit terms in sociological generalizations superfluous. Nonsense!

Constructs as a Type of "Characteristic Term"

A typology of terms in FTC should force recognition that some words/terms denoting quantitative characteristics may be far from empirically applicable, meaning that they do *not* designate something considered as being observable or measurable.[6] Such terms are here identified as *constructs*, meaning a word/term that a theorist may or may not define and, even if defined, the definition is *not* construed as (1) being complete and clear or (2) as pertaining to something measureable even though it clearly denotes a quantitative characteristic. Accordingly, "class conflict" in Postulate 1 is identified as a construct.

All such identification is a matter of the theorist's judgment, but should a critic question the designation of "class conflict" as a construct, he/she should *attempt* to frame a clear definition and a related measurement procedure that does not (1) tax credulity and (2) is neither arbitrary (as when class distinctions are defined in terms of specific income or wealth levels) nor inconsistent with the field's research resources, including published data. The problems in defining "a social class" alone have long defied solution, all the more when it comes to the goal of an empirically applicable definition (one known to assure agreement as to who is and who is not a member of the social class in question, such as "middle class").

Unfortunately, what has been said about "class conflict" extends to virtually all of sociology's major terms (e.g., social integration, social structure, power), which is to say that they promise insufficient empirical applicability. But a list of such terms would not fully disclose that sociology's vocabulary is a sea of metaphors. However, the point is not that sociologists should check their seemingly unquenchable taste for metaphors; rather, the metaphors can be identified as "constructs" and appear in theories only as components of axioms and/or postulates.

Concepts as a Type of "Characteristic Term"

Described succinctly, to identify a "characteristic term" as a concept is to claim that it is empirically applicable, at least sufficiently so. Hence, concepts are the opposite of constructs in two respects. First, when identifying a "characteristic term" as a concept and defining it, the theorist is claiming, tacitly or expressly, that the definition is complete and clear, something necessary for empirical applicability. Second, if a concept in a theory denotes a quantitative characteristic, its empirical application requires a formula and adherence to some procedure, including requisite kinds of data, all in the theory's extrinsic part.

In recent decades several studies of income inequality have been published, some of which relate to countries (for extensive references, see Rosanvallon 2013 and Clark 2013). Those studies justify designating "income inequality" as a concept, and a theorist might refer to them as instructions for researchers. Surely such references are the mark of a mature science, and ultimate maturity obtains when testable theories need not have an extrinsic part, meaning that the field's conventions are well established and inclusive. Despite a lengthy natural history stage, not even astronomy or physics is fully mature; and if only because sociology has never gone through even a remotely comparable stage, a *systematically testable* sociological theory without a lengthy and thorough extrinsic part is unimaginable.

As an aside, sociology's founding fathers (e.g., Comte and Spencer) set the field on the wrong course. Impressed by achievements in physics and biology (the theories of Newton and/or Darwin in particular), the fathers immediately set sail for grand theories, without recognizing the necessity of a long "natural history stage" in the development of a science, astronomy being the outstanding case. It is in the natural history stage that a science develops the terminology and procedures that are essential for formulating testable theories. Having not gone through a long natural history stage (despite the work of Adolph Quetelet and Frédéric Le Play), sociology became a hodgepodge of well-known but untestable theories.

Enough has been said to anticipate this objection: the advanced sciences get along, so to speak, without a MFTC. Anyone who so objects (1) has never read Newton's *Principia*, (2) is ignorant of the variety of conventional and empirically applicable terms in the advanced sciences—unit terms especially—that have nothing whatever to do with mathematics, and (3) does not recognize that unit terms are not explicit components of equations (one of the problems when stating a theory in the form of equations). Unfortunately, sociology is not a field wherein mathematics is a suitable MFTC. For that matter, the role of mathematics in biology is debatable, and even in physics the "string theory binge" may culminate in doubts about mathematics as being an entirely adequate MFTC.

Referentials as a Type of "Characteristic Term"

Recognition of a third type of "characteristic term," *referentials*, is truly peculiar to the proposed MFTC. Again, "income inequality" is identified as a concept; and for Postulate 1 to be a component of a systematically testable theory, in the theory's *intrinsic part* there would have to be something like:

Transformational Statement 1: Among countries, the greater the income inequality in a country, the greater RII for that country.

The "RII" is a referential pertaining to income inequality, one designating a computational formula and a description of the requisite kind of data, both in what would be the theory's extrinsic part.[7] Lest the concept/referential distinction appear redundant, for any concept there may be several alternative referentials; and the theorist's awful chore is to assess the alternatives with a view to promoting both the theory's testability and its predictive accuracy.

So a transformational statement links a concept and a referential, and it suggests a change in language, the introduction of some kind of metric. Why introduce that change? Because, regardless of the science, it is likely that for some quantitative terms there are alternative measurement procedures, including not just alternative formulas but distinctions as to kinds of data, perhaps even alternative methods for gathering the data, whatever the kind. Thus, in criminology, for several types of crime, the data for computing the rate may be in the form of official incidence statistics, victim-reported incidents, or self-reported incidents.

The suggestion is not that all alternative kinds of data are even somewhat equally reliable, a matter of the empirical applicability of the procedure for gathering or creating data. Rather, the theorist must choose among them; and *availability*—sometimes a matter of the field's research resources, sometimes a matter of a data source's willingness or ability to yield information—is crucial. So at least in sociology the creation of a formula frequently involves considerable complexities. Whereas the concept in a transformational statement pertains to a variable (e.g., the areal extent of a city, the frequency of burglary) without any stipulation as to metric or kinds of data, the corresponding referential denotes a formula for computing values and also kinds of data. But in explicating a referential, the theorist is concerned with both empirical applicability and the need to maximize the theory's predictive accuracy.

Why have two "characteristic terms" for what appears to be the same notion, in this case "income inequality" and "RII"? Answering all too briefly, regardless of the definition of "income inequality," there are various alternative formulas for computing the amount of it, in addition to diverse kinds of data. What is the most appropriate among the alternatives? Even ignoring the need to consider empirical applicability (always a matter of being sufficient or not), there may be no basis for a confident answer. Unfortunately, the theorist may be haunted by doubts to the point that in the extrinsic part of the theory he/she introduces alternative referentials, saying in effect "When testing the theory use the one that resources or other practical considerations dictate, and if possible apply both referentials to identify the one that results in the greatest predictive accuracy." If the reader recoils at such a prospect, bear in mind that, whatever the theory, "multiple referentials for the same concept" do not preclude test outcomes indicating *no* predictive accuracy at all regardless of the referential.

Even if there is only one referential for the concept in question, explication of it in the *intrinsic part* of the theory is not feasible, especially with a view to making that part comprehensible. Indeed, the failure to employ a MFTC is one of two principal reasons why reports of defensible tests of sociological theories are rare. Another reason is that inserting definitions and instructions in a theory's premises tends to confuse, very possibly to the point of discouraging attempts to test the theory. The general point: sociological theories commonly lack referentials, and their absence indicates that the theorist blithely assumed that researchers will meet the need, a sure path to controversy even if tests are conducted.

Without referentials, prospects for tests of the theory are drastically reduced, and even when tested, unconstructive debates about the outcomes are a near certainty. Nevertheless, it is virtually a sociological tradition for theorists to ignore measurement problems. Despite glaring contrasts in perspectives (e.g., functionalism vs. Marxism) the most common trait of famous sociological theorists (e.g., Talcott Parsons) is ignorance of or indifference to the field's research literature, a guide to empirical applicability.

Relational Terms

Consider what would be construed as a truly incoherent set of terms: among countries, class conflict, income inequality. The conjunction of the two "characteristics terms" is unintelligible because there is no *relational term* that links them.

Space limitations preclude an adequate treatment of relational terms; hence, note that an extensive treatment has been published (Gibbs 1994: 180, 192, 292–295, 359), and more will be said later in this chapter. At this point it suffices to say that only two relational terms are prescribed by the illustrative MFTC, "greater … greater" and "greater … less." The first of the two in Postulate 1 implies this untestable prediction: If a country's class conflict exceeds that of another country, then its income inequality also exceeds that of the other country.

Now consider the second premise:

Postulate 2: Among countries, the greater the class conflict in a country, the less the intergenerational occupational mobility in that country.

The relational term, "greater ... less," translates as this untestable prediction: if a country's class conflict exceeds that of another, then its intergenerational occupational mobility is *less* than that of the other.

Various observations can be made in support of Postulate 2, commencing with two contentions. First, class membership is suggested by occupational titles, such as "proprietor" and "laborer."[8] Second, intense class conflict is virtually certain to be manifested in laws and/or extralegal norms that prohibit or somehow lessen occupational choices for members of some classes, thereby *promoting* parent/child resemblances as regards occupation (hence *less* intergenerational occupational mobility).

Such plausible observations, though not systematic tests, may promote a theory's plausibility; but plausibility is not enough, though it hardly need be said that FTC scarcely denies the desirability of plausible assertions. Nonetheless, if theories are to be assessed strictly in terms of plausibility, why even attempt to test them? The question is commonly ignored by sociologists for one simple reason; they do not have a defensible answer.

But the question does throw some light on the seeming reluctance of sociologists to adopt FTC. They can argue that the use of a natural language discourages ridiculous or vacuous statements in the formulation of a theory. However, it is far from clear how FTC is conducive to ridiculous or vacuous statements, nor for that matter how the use of a natural language precludes such statements. The point is that the two modes of theory construction simply do not differ in that regard.

Now consider the third and last premise in the illustrative theory:

Postulate 3: Among countries, the greater the class conflict in a country, the greater the property crime rate of that country.

The second "characteristic term," property crime rate, is identified as a concept for three reasons. First, it can be defined clearly and completely as the average of the annual incidence rates for three types of crime: robbery, burglary, and theft. Second, there is a conventional procedure and formula for computing incidence rates. And, third, relevant data are published for several countries (until recently by the International Criminal Police Organization), and for present purposes the reliability/comparability of such data is irrelevant.

For some, Postulate 3 would not pass the plausibility test. However, if plausibility must be promoted, it will suffice to say that several publications provide numerous arguments that suggest support of Postulate 3 (see, especially, York and Clark 2006). It comes in reports of intense alienation, suggesting a level of anger and hatred among the proletariat in putative capitalist countries that is likely to be manifested in property crimes.

In Defense of the Relational Terms

An extensive defense of the prescribed relational terms has been published (Gibbs 1994). So it suffices to say that "ordinal relational terms" are much more realistic for sociology than are the principal alternatives, a causal language or equations. A causal language introduces irresolvable issues and insoluble problems, notably (1) the need for the necessary/sufficient distinction, (2) widespread differences of opinion as to evidence of a causal relation, and (3) all of David Hume's arguments (including versions by his disciples) about the very notion of causation.

As for equations, multivariate regression or not, even ignoring the need for explicit unit terms, they would call for a most *unrealistic* criterion in judging the predictive accuracy of sociological generalizations, that being predictions of precise values of variables. Yes, the formulation of sociological theories may develop eventually to the point that the component generalizations can be stated as equations, but we are not yet there; far from it.

There is a problem that goes unrecognized by staunch advocates of "mathematical sociology" (see, e.g., Fararo 1997). Equations pertain to the relation between variables denoted by "characteristic terms"; as such, they exclude explicit unit terms. In the advanced sciences perhaps the appropriate unit term—that which may maximize a generalization's predictive accuracy—need not be made explicit; but that is far from the case for sociology. Yet sociological theorists often appear oblivious to the point that the predictive accuracy of any sociological generalization is first and foremost contingent on the generalization's unit term.

The foregoing treatment of relational terms is all too brief, especially as regards the avoidance of a causal terminology and the use of equations. A more extensive treatment, issues in particular, would fragment the presentation of an illustrative theory all too much; so the subject is treated at greater length in the next chapter.

Residual Terms

Until sociology's conventions develop to the point that the field's generalizations can be stated as equations, purely syntactical words/terms (e.g., and, or, the) will be needed to state the intrinsic part of theories. So the proposed MFTC is not entirely divorced from natural language conventions, nor should it be.

Even if sociology should develop to the point that a mathematical language eliminates the need for *ordinal* relational terms, it is doubtful whether even physics can make do without extrinsic parts of theories or rely entirely on mathematics in stating the intrinsic parts. The latter fuels criticism of quantum theory, and excessive reliance on mathematics or inappropriate mathematics could seal the fate of string theories (see Unzicker 2013).

Types of Statements in the Intrinsic Part of a Theory

The foregoing typology of terms is fundamental in the proposed MFTC, as it is the basis for distinguishing axioms, postulates, propositions, and theorems in a theory's intrinsic part. Specifically, the constituent "characteristic terms" in a statement determines its type. That relation was suggested previously, and a more elaborate treatment is needed. So, reader, hang on.

Axioms

In the bivariate case an axiom asserts an empirical relation between two of an entity's characteristics, both designated by a construct. Even though constructs are not empirically applicable, an axiom is nonetheless a synthetic statement, one whose defensibility is determined by how it relates directly or indirectly to the world of events or things (some if not all neither observable nor measurable) and not simply by the meaning of the constituent words/terms.

Although the illustrative theory does not include axioms (you might say mercifully), one can be readily devised: among countries, the greater the relative power of the dominant class in a country, the greater the class conflict in that country. Whatever the country, the prospects of measuring "the relative power of the dominant class" may well be *less* than the prospects of measuring "class conflict." A measure of the "dominance" characteristic would require not only identification of dominant class members but also some measure of their relative power, collectively or individually. So, the first characteristic term in the illustrative axiom can be thought of as more of a construct than is the second term, which is to grant that empirical applicability is actually a matter of degree, though "none" or "maximum" is not precluded.

Despite what has been said about it, the illustrative axiom does assert an empirical relation, which is to say one not true or false by the very meaning of its constituent words/terms or syntax. However, if an axiom's constructs are not linked with concepts in the form of postulates, why would a theorist "tack on" such an axiom? He/she might do it to expand the theory's scope and/or in the reasonable hope that eventually someone will link all of the constructs with concepts through additional postulates. Nevertheless, "tacking on" an axiom does not further testability, and the price is a reduction in parsimony (the ratio of conclusions to premises). That consideration is obvious only if the theory is stated formally.

Postulates

While many philosophers, mathematicians, and scientists evidently equate axioms and postulates, we draw a distinction (yes, patient reader, another damn distinction). In the bivariate case, a postulate relates a construct and a concept.

At times the axiom/postulate distinction is difficult to maintain. It depends on the strength of the theorist's claim, tacitly or expressly, that her/his definitions of the concepts are complete, clear, and concomitantly empirically applicable (the latter also entails research resources and possibly other restraints). In any case, it is always a matter of the theorist's choice, though perhaps also her/his reputational funeral.

Three illustrations have been given in the way of Postulates 1, 2, and 3, each connecting class conflict with income inequality, intergenerational occupational mobility, or the property crime rate. Observe again that each postulate asserts an empirical relation between a characteristic designated by a construct and a characteristic designated by a concept. Then comes the rub. Assertions made via either an axiom or a postulate are not testable in any direct sense. Even so, were the asserted relations testable, we might find gross errors in predictions despite no measurement errors; and being "theoretical," there are no measurement errors in the case of axioms or postulates.

Additional Transformational Statements

Two more transformational statements are needed along with Transformational Statement 1 (*supra*) as penultimate steps to complete the illustrative theory. Both instances include referentials not previously considered. Their inclusion is necessary because without them the connection of the concepts with observational data would be obscure at best.

Transformational Statement 2: Among countries, the greater the intergenerational occupational mobility in a country, the greater the ROM for that country.

Transformational Statement 3: Among countries, the greater the property crime rate for a country, the greater the RPCR for that country.

As before, each referential designates a computational formula and the requisite kind of data, both of which would be in the theory's extrinsic part.

Now we are getting closer to prospective observations on the predictive accuracy of the theory. Still, problems remain. How do we go about creating or selecting the appropriate formula for computing a value that expresses the amount of what is designated by a concept? The question is all the more difficult because of the sheer size of each population (residents of a country), and the always severely limited research resources in sociology make it virtually imperative to stipulate the requisite kind of data in published form (e.g., census reports).

As for the ROM formula, there is an advantage in that reports of comparative studies of intergenerational occupational mobility are in the literature, some at the international level (e.g., Beller and Hout 2006; Wong 1990). So the theorist may identify such an international study and even prescribe use of the reported mobility values for countries.

The foregoing references to published studies of intergenerational occupational mobility reflect recognition that referentials, whatever they are, must be sufficiently empirically applicable (i.e., hopefully to give the theory greater predictive accuracy than that of any contender). Accordingly, with a view to making use of the findings of more than one study, the theorist can further the testability of her/his theory by formulating more than one version of the intergenerational occupational mobility referential. The theorist is likely to find the choices narrow and difficult, but the limited research resources of sociology may force a sacrifice of confidence for testability.

Returning to Transformational Statement 3, the referential, RPCR, would designate some formula and data instructions for computing a country's property crime rate. Virtually everything said about the ROM referential extends to RPCR. However, there is a conventional formula for computing crime rates, and the one in question differs from the conventional only in that it would be the average of three incidence rates—for robbery, burglary, and theft.

As indicated earlier, until recently rates for various types of crimes in several countries were published by an international organization, and the concern of the theorist with empirical applicability may lead her/him to prescribe the use of rates published by national agencies (e.g., the F.B.I. in the case of the U.S.). There are justified concerns about both the comparability and reliability of published crime rates, but in a field sorely pressed for research resources, as sociology definitely is, published data are better than none at all. Yet all too often the choice is between one of two sacrifices—testability or predictive accuracy.

Theorems and the "Sign Rule"

In keeping with convention, a theorem is a statement deduced from a theory's premises, but further clarification is needed. In the proposed MFTC all of the "characteristic terms" in a theorem are concepts, and each theorem is deduced in accordance with the sign rule. In the illustrative theory all theorems are deduced from postulates, though in other theories some or all theorems might be deduced from combinations of axioms, postulates, and/or propositions (defined subsequently).

The sign rule is best explicated in terms of three official requirements. First, when the relational term is "greater ... greater," the premise is assigned a positive sign (+); and when "greater ... less" the assigned sign is negative (−). Second, deduction of a theorem requires a series of two or more premises that have a "unit term" and a "characteristic term" in common. Third, given such a series, the deduced theorem has the "greater ... less" relational term *if and only if* there is an *odd* number of such relational terms in the premises.

To better understand the sign rule, consider **Theorem 1** as deduced from Postulates 1 and 2: **Among countries, the greater the intergenerational occupational mobility within a country, the less the income inequality in that country.** The characteristic term "class conflict" appears in both postulates (1 and 2), and there is an odd number of premises in which the relational term has a negative sign (i.e., "greater ... less"), the *one* instance being Postulate 2. It follows that the deduced theorem has a negative sign, which translates as "greater ... less."

The sign rule applies whatever the "characteristic terms" in the premises (constructs, concepts, or referentials); and there is no restriction as to the type of statement that can be deduced, though any type other than a theorem should be further identified as "implied." So it is possible to have a theory in which there are two or more axioms (all "characteristic terms" are constructs), and they imply at least one other axiom. Throughout this book, to simplify, "derived," "deduced," and "implied" are treated as equivalents. So to say that some statement is "implied" by others is to say that it is deduced or derived from those other statements.

Application of the sign rule to the three postulates in the illustrative theory results in two additional theorems, as follow, along with a repetition of Theorem 1.

Theorem 1 (from Postulates 1 and 2): Among countries, the greater the intergenerational occupational mobility within a country, the less the income inequality in that country.

Theorem 2 (from Postulates 2 and 3): Among countries, the greater the intergenerational occupational mobility in a country, the less the property crime rate for that country.

Theorem 3 (from Postulates 1 and 3): Among countries, the greater the income inequality within a country, the greater the property crime rate for that country.

Testable Theorems

Because the "characteristic terms" of a theorem are concepts, the theorem is considered testable. That point is consistent with convention, but another step is necessary to realize predictions about the association of actual values of variables. Such generalizations are deduced from transformational statements along with the theory's other premises (here, three postulates). They are identified as *systematically testable theorems*, and there are three in the illustrative theory, as follow:

Systematically Testable Theorem 1 (from Postulates 1 and 2 along with Transformational Statements 1 and 2): Among countries, the greater ROM for a country, the less the RII for that country.

Systematically Testable Theorem 2 (from Postulates 2 and 3 along with Transformational Statements 2 and 3): Among countries, the greater the ROM for a country, the less the RPCR for that country.

Systematically Testable Theorem 3 (from Postulates 1 and 3 along with Transformational Statements 1 and 3): Among countries, the greater the RII for a country, the greater the RCPR for that country.

A Digression

There are serious issues concerning the sign rule and the very idea of testing a theory through deductions, but they are best confronted in a later chapter after treating some other subjects. However, before these treatments, consider a strategic merit of the proposed mode.

The mode makes it possible to identify *particular* premises as "supported" or "unsupported" even though no premise in the series is testable. To illustrate, suppose that tests of the illustrative theory support *only* Testable Theorem 1, a likely outcome in light of two studies cited by Wilkinson and Pickett (2009: 495). How could such an outcome possibly indicate anything other than support for Postulates 1 and 2 along with Transformational Statements 1 and 2 but *not* for Postulate 3 and Transformational Statement 3? After all, *only* Postulate 3 and Transformational Statement 3 would have entered into the derivation of the two *unsupported* testable theorems (2 and 3). The forthcoming diagram of the illustrative theory will clearly reveal the logic of the inference.

Temporal Quantifiers: A Final Complexity

In keeping with the conventional form of generalizations in sociology, the postulates, theorems, or testable theorems in the illustrative theory suggest nothing specific as to the temporal relation between the variables. Yet it would tax credulity to assume that the asserted bivariate associations are *not* contingent on the kind of temporal relation between the variables. Thus, contemplate an increase in class conflict (CC) in some country. Surely Postulate 2 would anticipate a decrease in occupational mobility (here, always intergenerational) though not instantaneously, but the postulate does not even suggest the period during which the CC change would be most fully expressed.

Given that CC is not measurable and occupational mobility is measurable only through the transformational link with its referential (ROM), the *temporal quality* of ROM and each of the other two referentials, RII and RPCR, is crucial. Unless that quality is stipulated, investigators will not know how to compute *referents* of RII, ROM, and RPCR (i.e., quantitative values of those referentials). Thus, in the case of RPCR, in each of the two testable theorems the temporal quantifier *could be* "RPCR at T_5," indicating that the theorist thought of CC at the beginning of *any year* as most fully expressed by the annual property crime rate five years later. But if it is, say, "RPCR at T_3" then the annual rate must be that for a period three years later. Note that the temporal quality for any referential is relative to that of CC, which is set at, so to speak, "T_0" (though any particular calendar year could be an instance). Accordingly, if the *temporal quantifiers* are "RII at T_1," "ROM at T_5," and "RPCR at T_4," then the theorist has thought of CC at the beginning of *any year* as most fully expressed by RII one year later, by ROM five years later, and by RPCR four years later.

Although previous examples pertain to static values, the theorist may think of CC and all other variables (concepts or referentials) in terms of change, absolute or proportionate. In the *absolute* case the "temporal quantifier" of a referential, *R*, would be expressed by linked numbers, as for example "R_{5-2}", in which case *R* is the referential for some particular variable (e.g., income inequality), and the reference is to the absolute amount of change in the referents (values) between two time periods, one three years later than the other (proportionate change would be signified by "$R_{5/2}$").

The point/period distinction is needed if only to indicate that the prescribed language is not restricted to years, let alone particular historical years (e.g., 2010, 2015). Rather, the numbers in temporal quantifiers may be nanoseconds, seconds, minutes, months, years, or even centuries. Finally, a referent may be the average of two or more points or periods. Thus, "$R_{[1-3]}$" could mean an average over three points or periods, as when, for example, the crude birth rate of some entity is the average of the 2009, 2010, and 2011 annual rates.

Although a theorist is obliged to clarify the temporal quantifiers, he/she is free to use whatever symbols he/she prefers. He/she is also free to insert temporal quantifiers in the premises, the testable theorems in particular. However, since the symbols must be clarified *in the theory's extrinsic part*, the theorist might as well stipulate the temporal quantifiers and clarify them when defining the referential in question (i.e., they need not be in the premises). Whatever, a failure to stipulate and clarify invites idiosyncratic tests, if any at all, of the theory. If the theorist does not stipulate temporal quantifiers, it will fall to those who attempt to test the theory. Finally, an additional and more elaborate explication of temporal quantifiers has been published previously (Gibbs 1994, especially 203–206 and 233–235).

Propositions

A proposition is a statement that asserts some kind of empirical relation between properties or variables designated by concepts. There are no propositions in the illustrative theory, but an illustration has been given previously. It is here repeated: among countries, the greater the degree of urbanization in a country at any point during a year, the greater the suicide rate of the country for that year.

The statement is identified as a proposition because it relates two concepts. But it was not deduced from any premises; hence, its theoretical significance for sociology is obscure.[9] For that matter, it is extremely difficult to find generalizations in the sociological literature that are explicitly identified as propositions *in light of some explicit typology of statements* (as here) and clearly of theoretical significance. By contrast, physics is a study in such propositions, though commonly designated as "laws." That word is not used here to designate a type of statement, the reason being the connotation of "no known exceptions." That connotation makes the notion alien for many sciences—perhaps all of them—and certainly sociology.

The phrase "no known exceptions" pertains to predictive accuracy, something that does not enter into the present typology of statements. Nonetheless, in all sciences the word "proposition" suggests a generalization that applies to an infinite category of events or things (i.e., no spatial and/or no temporal limits). That connotation may be the primary reason why contemporary sociologists rarely speak of propositions. If so, it is most unfortunate, because in any science it is uniformities that pose the greatest challenge for theorists, meaning that they formulate theories with a view to deducing known or assumed uniformities. Stated in the terminology of the proposed MFTC, a *deduced* uniformity is identified as a theorem, but it can be thought of as an *implied* proposition. Accordingly, all three of the theorems in the illustrative theory can be thought of as *implied* propositions.

In passing note again that for some obscure reason sociologists are prone to refrain from the use of the word "proposition" to identify an assertion. By contrast, the compilation of "sociological principles" has been a beehive for generations, the latest being Jonathan Turner's work (2010). Although admirable for various reasons, Turner's compilations are not based on any explicit typology of terms (e.g., akin to what has been identified as unit, relational, or characteristic), and the logical relations among his "principles" are generally obscure. There is more on Turner in a later chapter.

Tests and Formal Theory Construction

All of the previously recognized types of terms and types of statements appear in Figure 6.1, a diagram of the illustrative theory. Observe that the diagram includes "hypotheses." They are treated in connection with "tests," the subject to which we now turn.

There are rare instances where the writer has treated "theory" and "hypothesis" as interchangeable words (K. Parsons 2014: 56). The rarity is understandable because systematic tests of a theory require deductions of hypotheses from the theory's theorems. That terminology is adopted here, but note that use of the proposed MFTC in stating a theory ends with the

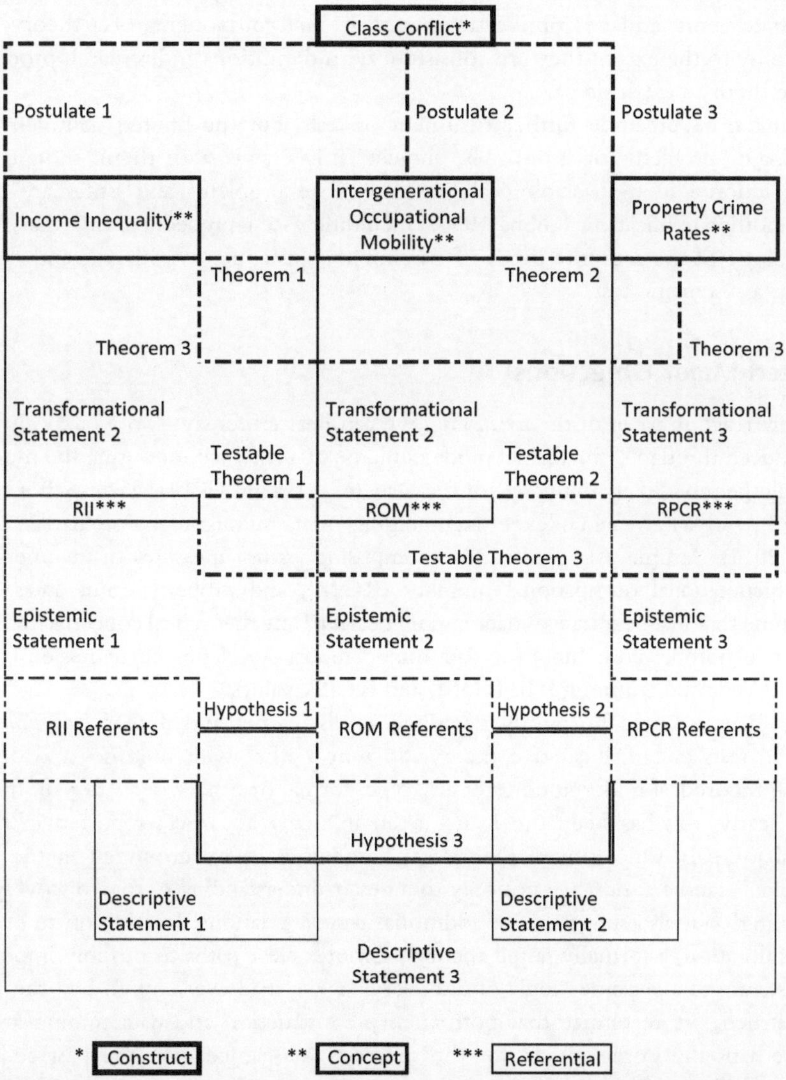

FIGURE 6.1 Diagram of the Illustrative Theory about Relations among Countries and Test Components

deduction of *at least one* testable theorem (i.e., hypotheses are deduced only in tests of the theory, not in statements of it). Accordingly, a theorem or any isolated proposition is tested only through tests of hypotheses that have been deduced from it.

The immediate implication is that hypotheses are not components of a theory, whether the theory is quantitative or qualitative.[10] Rather, in the case of a quantitative theory a hypothesis is a prediction about *particular* sets of values of variables, a prediction deduced from a theory through a testable theorem and "epistemic statements." Each epistemic statement links a referential with values identified as "referents" in a table (truncated when there are so many entities that presentation of all values is not feasible in reporting a test). So a test is essentially a "descriptive statement" about particular things—values in the case of a quantitative theory—a statement that is either consistent or inconsistent with a hypothesis that has been deduced from the testable theorems and epistemic statements (for examples of such "test terminology," see Gibbs 1994: 216–219 and 339–343). Although hypotheses, the epistemic statements, and descriptive statements are *not* components of a theory, they are defensible only to the extent they are consistent with definitions and research procedures set forth in the theory's extrinsic part.

Space limitations preclude further treatment of tests, but the limited treatment does *not* mean no major problems or issues. The subject fairly crawls with them, though they are commonly ignored in the sociological literature. The problems and issues are treated at length in another publication (Gibbs 1994), including a test procedure, one that includes a special typology of the requisite kinds of statements (epistemic, hypotheses, and descriptive) in reporting a systematic test.

Anticipated Major Objections

As one likely reaction to all of the foregoing, we can hear critics say: "So what? Any empirical relation deduced through some formal mode could be *discovered* without using the mode." That reaction is all the more likely in this age of Big Data (e.g., Baesens 2014, Mayer-Schönberger and Cukier 2013), with a new field of expertise emerging—data mining (e.g., Foreman 2013, Witten and Frank 2011). Assume a large data base comprising *asserted* measures of income inequality (RII), intergenerational occupational mobility (ROM), and property crime rates (RPCR). Further assume that in an exercise of data mining bereft of any theoretical concerns (referred to in bygone days as number crunching) the data miners report correlation or regression coefficients pertaining to associations among RII, ROM, and RCPR values.

Such coefficients could be rightly described as discoveries and perhaps *identical* to those computed in tests of the illustrative theory. So why bother with premises and deductions, indeed with theoretical underpinnings of any type, formal or otherwise? After all, findings are findings. Clearly, this has been the route taken in many an article reported in the major sociological journals. Our answer: theoretical frameworks, here grounded in the construct "class conflict," can be conducive not only to a better understanding of the "discovery findings in data mining" but also suggestive of additional *possible* relations. In addition to serving as a map for exploration, a formally stated theory promotes clear paths to possibly important discoveries (as close to a genuine "logic of discovery" as we shall ever come). Lest the foregoing be misconstrued, we recognize that both theory construction and data mining are worthy pursuits, perhaps the prime reason for the successes of science over the past century; but success is most likely when the two activities are somehow integrated.

Finally, suppose the illustrative theory is eventually tested, and the reported correlation coefficients are far from impressive, indicating that the predictive accuracy of the theory is negligible, taken by itself or relative to other theories. The immediate point is that no MFTC can possibly guarantee even appreciable predictive accuracy, whether in absolute or relative terms.

The primary purpose of a MFTC is to promote systematic tests of theories, not to assure predictive accuracy. Without tests, there can be no truly rational decisions about a theory's future. If tests indicate grossly inaccurate predictions, the theorist should seriously consider "packing it in." The same is true if the tests reveal far less predictive accuracy does than some contending theory, though a synthesis is a likely possibility. If there are no contenders and test outcomes indicate levels of predictive accuracy beyond chance expectations, the scene would be set to expand the theory by adding variables that specify conditions in which the anticipated associations will increase. Such specifications would necessarily be assertions, testable through the "set comparison method" or some multivariate statistical technique.

Prospects for Expansion of the Illustrative Theory

The foregoing should not be construed as limiting modification of the illustrative theory to attempts at increasing its predictive accuracy in light of tests. In particular, there are ample opportunities for expanding the theory's scope. Although such expansion should be the product of careful thought, it can be accomplished by a very simple step—formulating additional assertions about correlates of "class conflict," either as axioms or more postulates.

An expansion in the way of an axiom has been stated previously. Addition of a possible postulate comes to mind readily: among countries, the greater the CC in a country, the greater the "political violence death rate" of that country. Given that CC is a construct, the generalization's identification as a postulate implies that "political violence death rate" must be a concept. Justification comes from Taylor and Jodice's study (1983), wherein they reported a count of deaths from political violence in 136 countries.

Additional Features of Formal Theory Construction

Even cursory examination of a single issue of a sociology journal will suggest that sociologists study an astonishing variety of subjects. That variety does more than exacerbate the difficulty of delimiting the field's subject matter; it also forces recognition of a daunting problem, that of achieving even moderate coherence. Stated in the vernacular: how can sociologists get it all together?

Integrating Theories

Assuming that "coherence by fiat" is neither feasible nor desirable, there is only one obvious alternative—make theories the touchstone when delimiting sociology's subject matter. Briefly, if some kind of human behavior or social phenomenon is not treated in at least one accepted sociological theory, then it is outside sociology's subject matter, though tentatively. But the problem is not just the lacuna of accepted sociological theories. Additionally, even if accepted theories were abundant, the crucial "coherence question" would be something like: wherein lies their unity?

The question points an indisputable goal in any science—the integration or synthesis of theories. Fortunately, a fairly clear criterion of *successful* theoretical integration can be stated readily. Two or more theories have been successfully integrated when and only when it is possible to deduce a *unique* theorem from some combination of all of the premises. Unfortunately, what with the conventional mode of theory construction in sociology, even when dealing with only two theories, the deduction of a unique theorem is more a fond hope than an accomplishment. When the premises of conventionally stated sociological theories are combined, it is a hopeless muddle, a semantic swamp, even assuming confident identification of the premises (lots of luck). There are several problems, but FTC reduces or avoids most of them.

It is pointless to so much as contemplate the integration of theories that do not have logically related unit terms (literally the same unit term being the most obvious). Unless theories have a common unit term or otherwise logically related unit terms, then attempting to integrate them is to ignore the bromidic proscription of mixing apples and oranges. However, bear in mind that "class inclusion" is a logical relation; for that matter, also bear in mind that bromides are scarcely rules of logic. Thus, to repeat a previous point another way, if the unit term is "Among human populations," the theory's *range* is enormous. Among the numerous possibilities, that unit term would create the *potential* for integrating the theory with another in which the unit term is "Among countries," "Among cities," or even "Among human age groups." After all, the residents of a country or a city surely constitute a human population, as do members of some particular age group. However, those illustrations do not reveal the complexities, all pertaining to a fully developed "logic of inclusion," one that would preclude equating "population of a country" with "combined populations of all cities in a country," let alone either one with "population of a particular age group." Indeed, mention of the latter brings to mind the crucial difference between two possible unit terms, "Among human age groups" vs. "Among human age groups in the same country." The most general point is that when it comes to unit terms and a logic of inclusion, the conventions of a natural language simply will not do, one of the main reasons why those conventions are not sufficient for a mode of theory construction.

The point just made about conventions suggests something about the use of a causal language in the way of relational terms. Consider a series of generalizations, all in abstract form *other than* the relational terms. First, among *U*s, variation in *X* is *the* cause of variation in *Y*. Second, among *U*s, variation in *X* is *a* cause of variation in *Y*. Third, among *U*'s, variation in *X* is the *direct* cause of variation in *Y*. Fourth, among *U*s, variation in *X* is the *indirect* cause of variation in *Y*. Fifth, among *U*s, an increase in *X* is a *necessary* cause of an increase in *Y*. Sixth, among *U*s, an increase in *X* is a *sufficient* cause of an increase in *Y*. Even though the six generalizations have identical unit terms and identical characteristic terms (here variables), their relational terms are quite different; indeed, different to the point of making the pairs of generalizations contradictory (i.e., "the cause" vs. "a cause"). Hopefully, the reader will see that all such problems are avoided by the relational terms, "greater … greater" and "greater … less" in the proposed MFTC.

The final integration requirement may well be the most obvious. Stated as briefly as possible, generalizations or theories are *potentially* integrative if they have the same unit term and at least one "characteristic term" is a component of all of them. The shared "characteristic term," be it a construct or concept, denotes a variable. Unfortunately, there are two crucial contingencies, both pertaining to the extrinsic parts of the theories. First, definitions of the "shared characteristic term" must be congruent, an irrelevant requirement for *undefined* constructs; hence, their use facilitates the integration of theories. Second, if the "shared characteristic term" is a quantitative concept, measurement instructions must be at least approximately the same.

Each integration requirement is difficult to satisfy when the theories are stated in some natural language. The immediate problem is the bewildering variety of *possible* unit terms, relational terms, and "characteristic terms" in what purports to be one theory. So it is no wonder that even attempts at the integration of sociological theories are rare. For the most part, sociological theories simply sit there in splendid isolation. However, the point is not just that sociological theories commonly defy integration. Additionally, whatever the science, if two or more theories are stated in some natural language, any attempt at integrating them is likely to be an embarrassing failure. Even so, no claim is made here as to the utility of the proposed mode other than its application to theories about quantitative sociological variables.

The Importance of Unit Terms

One more benefit of FTC awaits identification. It is perhaps less important than integration of theories but certainly of second rank.

Although not heretofore stressed sufficiently, the importance of a theory's unit terms defies exaggeration. It is not just a matter of their empirical applicability, though that is the immediate consideration. No less important, there is always the awful possibility that negligible predictive accuracy in tests of the theory is due largely to an inappropriate unit term. To illustrate, suppose that tests of a theory indicate negligible predictive accuracy for *all* theorems. What could account for such a disaster? It would not likely be something about all of the variables, such as gross measurement error (really a matter of empirical applicability) in all referents. There would be two more likely possibilities.

The Unit Term's Empirical Applicability

Whatever the theory, maximum empirical applicability of the unit term (UT) is a *necessary* condition for maximum predictive accuracy in tests of the theory. That point takes on additional significance when there are multiple theorems. Consequently, if tests indicate negligible predictive accuracy for *all* theorems, it could be due solely to the insufficient empirical applicability of the UT.

There are two possibilities as regards the kind of insufficient empirical applicability in question. They are not mutually exclusive, and both are all too likely in sociology. Recall that a theory's UT identifies the kind of entity that is relevant in testing a theory. So the most blatant case of insufficient empirical applicability would be where in testing a theory the investigators gathered data on the "wrong" kind of entities, meaning entities that do not satisfy the criteria stipulated by the theorist. For example, suppose that the UT in some theory is "primary groups," but for one reason or another (perhaps limited resources) in testing the theory an investigator gathers data on large classes in a huge high school, thereby ignoring the theorist's definition of a primary group, one that makes frequent interaction between any two members a necessary feature (i.e., by definition) of a primary group.

Unfortunately, there is no basis for even estimating the frequency with which the "wrong" entities are selected in tests of sociological theories, much less the impact of such mistakes. For that matter, you can search sociology journals endlessly without finding report of research on the empirical applicability of any sociological concept, UT or otherwise. As noted elsewhere, there are a few items in the sociological literature on social class terminology with some attention to empirical applicability, but very little more than that.

Inappropriate Unit Terms

Recall that we are considering possible reasons for *uniformly* negligible predictive accuracy in tests of multiple theorems in a theory. One possibility has really nothing to do with the empirical applicability of the theory's UT.

Suppose that a theory asserts some kind of relation *among cities* between a wide variety of pairs of variables (e.g., infant mortality and income inequality, unemployment rates, and suicide rates). So, if stated formally, there would be several theorems. Now suppose that the test of each theorem clearly indicates negligible predictive accuracy.

Observe that if stated in accordance with the proposed mode of theory construction, the theory's UT would be "Among cities." Hence, the question: What would have been the test outcomes had the theory's UT been "For any city over time"? There is no reason whatever to deny the *possibility* that the test outcomes would have been quite different (i.e., much greater predictive accuracy). And, if so, it would be obvious that the theorist selected an inappropriate UT.

As in the case of empirical applicability, there is no basis for even estimating the frequency with which sociologists stipulate an inappropriate UT in formulating a theory, and the reason is the same—all too rare reports of research on the subject. However, the subject is more closely linked with FTC than is empirical applicability. After all, the empirical applicability of some term is not increased by its use in a formal theory, but in using a formal mode to state a theory the theorist is much more likely to make a *careful choice* in selecting the UT (i.e., greater awareness that the wrong choice would doom the theory). By contrast, the use of nothing more than the conventions of a natural language in stating a theory gives the theorist a license to use a bewildering variety of possible UTs (e.g., societies, groups, populations, persons, cultures, civilizations, associations, organizations). Whatever the theory, how could all such UTs be appropriate?

And if you doubt uncritical choices of UTs, read Emile Durkheim's *The Division of Labor in Society* (1949 [1893]). Durkheim is known for having set forth a theory in that book, and given the book's title one would anticipate a focus on some particular type of entity (if not "society," then something else). But no, what we find is an indescribable variety of UTs. However, what is said here about Durkheim's theory applies to all of the "grand theories" (read K. Marx, V. Pareto, H. Spencer, M. Weber or any other "theorist" in sociology's pantheon), and there is a glaring common denominator. Any "grand theory" is set forth in hundreds of pages in which there is a multitude of UTs. Consequently, how can anyone say with confidence that they know what the theory really is? Will sociologists ever cease warming their hands over ancestral bones? Let us hope so.

A Few Final Words on Formal Theory Construction

Exhausted readers will be relieved on coming to know that we are nearing the end of this chapter. Nonetheless, were it not for recognition of that exhaustion and space limitations, the chapter would go on and on. However, we must close with a few final words.

The history of any science is one of a very definite pattern. It always commences with a search for uniformities in some particular category of events or things, and later theories are devised with a view to deducing some of the discovered uniformities and perhaps others (deduction of heretofore undetected uniformities is the touchstone of a really impressive theory). The "searching-for-uniformities-period" commences in what is known as the "natural history stage," but neither really ever ends; and for that reason alone the "theory factory" never closes, though in sociology during recent decades closure appears likely.

But something went dreadfully wrong in sociology's history. It commenced with the totally premature stretch for theories, premature in that it commenced long before there was firm evidence of uniformities in the field's subject matter; and the very notion of a uniformity implies testability. Perhaps equally important, early prominent sociologists were much more ideologues than "searchers for uniformities." As cases in point, it would be startling should anyone deny that Comte was obsessed with the promotion of human welfare, that Spencer doggedly defended laissez-faire policies, that Durkheim was conservative to the point of writing as though the French Revolution never took place, that Pareto promoted fascist ideas, that Weber pursued liberal causes, or that Marx longed to bring down capitalism.

Has the promotion of ideologies by sociologists ended? By no means! Joe Feagin's presidential address (2001) to the American Sociological Association is only one of numerous continuations. Indeed, the promotion of ideologies has apparently escalated in recent decades. The paradox is that the first scourge of sociology—the preoccupation with "grand theory"—actually declined long ago, only to be followed by a decline of the interest of sociologists in theories, grand or otherwise (some evidence of the latter decline is introduced in Chapter 8).

The apparent decline of interest is unfortunate if only because an escalation could stimulate recognition by sociologists of the importance of uniformities in science, at least the quest for them. Again, a truly impressive theory is one conducive to systematic deductions of known or heretofore undetected uniformities, the very thing that is most likely to promote the search for uniformities and, let it be said, a sense of excitement in the field, something missing in sociology for decades.

The suggestion here is that in sociology the connection between theories and uniformities will not be furthered until sociologists adopt some MFTC. The promotion of testable theories through adoption of some formal mode would alone stimulate the search for uniformities; and the stimulation would be even greater by the deduction of heretofore unrecognized uniformities, something most likely through the integration of theories or generalizations. However, sociologists are sadly mistaken if they believe that systematically testable theories, let alone integrative theories, can be formulated without adoption of some MFTC.

Notes

1 For that matter, there is no rule in a natural language that calls for identification of statements as being either analytic or synthetic, much less one that calls for typification of synthetic statements (e.g., axiom, postulate, proposition, or theorem). If a naysayer doubts the need for such distinctions, he/she has never struggled with a theory formulated by Durkheim, Marx, Pareto, Simmel, or Weber, especially attempting to identify testable statements.

2 The problems and issues posed by formal theory construction can be best described by examining a particular mode. A lengthier version of the mode has been published (Gibbs 1994: 279–364); and it will be referred to subsequently, where space limitations preclude adequate treatment of some features of the illustrative mode.

3 Of course, the theorist may believe that the association between class conflict and income inequality holds somewhat the same for two or more kinds of entities, in which case he/she should identify each kind (e.g., "Among countries and cities"). So doing would not only clarify but also increase the theory's range, a dimension of predictive power (Chapter 4).

4 Despite the emphasis on "relative predictive power," it is admitted that there must be an alternative criterion when there is no contending theory. In that case, *absolute* predictive power is the only alternative to such criteria as plausibility, elegance, and intellectual satisfaction. As to *minimum* absolute predictive accuracy, the only solution is "exceeding chance expectation by at least one generally accepted criterion."

5 The maximum range for a sociological theory *could be* realized by this unit term: "Among populations and for any population over time." The emphasis indicates that the theory's range really depends on how "a population" is defined in the theory's extrinsic part, some possibilities being residents of any territorial political entity (country, state, city, etc.), members of some particular organization or association, or any kind of group (perhaps including experimental or situational groups). Should it appear that a theorist will select unit terms solely with a view to exceeding the range of contending theories, in so doing he/she will have jeopardized the theory's predictive accuracy, though perhaps expanding its testability.

6 Observability is scarcely relevant in the social sciences. Thus, we do not actually observe a wife or even a home owner; rather, we apply such words/terms on the basis of inferences. Nonetheless, in judging empirical applicability, the theory's UT is a crucial contingency. Thus, the empirical applicability of "frequency of interaction" is one thing when the UT is "among primary groups" but quite different when "among countries." In the latter case there is no realistic possibility other than identifying "frequency of interaction" as construct. Without doubt, such considerations test the patience of some sociologists. As a case in point, one of the authors knew a University of Arizona sociologist who invested months in mastering "path analysis" but balked at the complexity of formal theory construction.

7 Some readers may equate referentials with operational definitions. We question the very notion of an operational definition, at least in the case of sociology. To illustrate, if a particular social class is defined in terms of income or wealth, what is "operational" about the definition? Taking the question seriously, at least in sociology the term "operational definition" simply translates as "empirically applicable," at least in the sense of sufficiently.

8 At least in some populations the point applies to "caste membership." For that matter, the illustrative theory could be extended to "caste conflict" along with "class conflict," and it may well be that in contemporary countries "caste" (including race) is much more empirically applicable than is "class." Even so, "caste conflict" would be identified as a construct. The more general point pertains to conflict theory in sociology. Whatever its merits, advocates of that theory (more accurately a perspective) have yet to formulate a clear and complete definition of "conflict" at the aggregate or collective level, let alone a measurement procedure that even appears to offer appreciable empirical applicability. To repeat a previous contention, when formulating a theory about the "time dynamics" of conflict, Randall Collins (2012) never came close to recognizing conceptual problems and issues in defining conflict. Moreover, defining conflict is only a step toward defining class conflict, a subclass of an even more abstract notion long alluded to by sociologists, "social conflict" (see, especially, Coser 1956).

9 In any science a proposition can be important for various reasons, the most likely possibility being its position in a theory as a premise. However, what was an isolated proposition may be eventually deduced from the premises of a theory, in which case it becomes a theorem. Even though no premise in the illustrative theory is a proposition, note again that a proposition can be a premise. For that matter, a theory could be such that all of its premises are propositions (i.e., no axioms or postulates).

10 Most of this book pertains to quantitative theories. However, that focus should not be construed as a denial of the possibility or importance of qualitative theories, even instances having substantial predictive power. Furthermore, there is a real need in sociology for another mode of formal theory construction, one for stating *qualitative* theories (see Timmermans and Tavory 2012). There is even perhaps a greater need for a mode designed to state theories in which only some of the constructs and/or some of the concepts are quantitative.

7

MORE ON ISSUES AND PROBLEMS CONCERNING FTC

With the basic features of the proposed mode of formal theory construction (FTC) hopefully in your grasp, we turn to a more detailed treatment of issues/problems, or an introduction of new ones. Even though focused on FTC, this chapter, along with the previous, is a far cry from a complete treatment of the subject, not even if limited to sociology. Various issues/problems will be treated superficially, if at all, and subsequent attempts at clarification are themselves severely constrained by space limitations. Nevertheless, further awareness of issue/problems pertaining to FTC is within reach, and that is our aim. We begin with a subject closely related to the deduction of theorems.

Logic and Theory Construction

Daniel Bonevac (1990) begins his widely read and respected treatment of logic by describing it as "the study of correct reasoning." Seems simple enough. But 503 pages later, the reader, if paying attention, has encountered—among a host of other topics—truth trees, fuzzy logic, fallacies, counterfactuals, good arguments gone bad, quantifiers, and modality. Most relevant for our purposes, Bonevac examines "logically valid but empirically unsound arguments." So the conclusion: When dealing with logic, things are frequently not as simple as they first appear, not even nearly.

An alternative to the "logical" might be the mathematical representation of assertions. It would be naïve, however, to assume that a mathematical mode of theory construction would circumvent the issues. The question is not whether mathematics could be useful for abstract representation of the links among or between constructs, concepts, and referentials. Clearly it would be. Equations are actually essential when it comes to predicting specific values, and surely a necessary step when assessing contending theories in light of their relative "precisional accuracy." However, as previous observations on predictive accuracy have indicated, the "precisional" is only one accuracy criterion. Moreover, decades of development of conventions and research findings may be necessary before "precisional accuracy" becomes a *realistic* criterion in assessing sociological generalizations or theories. We are simply not there yet.

The Sign Rule Reconsidered

Not a great deal of thought is required to recognize that the issues and problems in contemplating formal theory construction go far beyond the prediction of precise values of variables. Indeed, they go far beyond the use of the "sign rule," although that use is the dominant issues in what became a substantial literature. As a guide to that literature, see references by Alessio (1982) to the publications of K.D. Bailey, H.M. Blalock, H.L. Costner, O.D. Duncan, R.K. Leik, S. Turner, and W.C. Wilcox. In passing, note that the literature on the sign rule effectively but strangely ended with Alessio's paper. Also note that Alessio and all of those authors he cited are clearly critics of the use of the sign rule in stating theories, at least sociological theories.

The most remarkable feature of the criticism is that with one exception (*infra*) the opponents of the sign rule ("use of" is henceforth understood) never came close to clearly identifying an alternative, much less something remotely resembling a mode of theory construction other than the conventions of a natural language. There is nothing more than the suggestion that mathematics provides a sufficient mode of theory construction for sociology. So let us eventually examine that suggestion, but for the moment bear in mind that the subject is only one of a multitude of issues when considering the central question: what is the most appropriate mode of theory construction for sociology?

The question forces recognition that any proposed mode is a form of logic; but that recognition poses a question, one that the critics of the sign rule never face, even though they use the word "logic" often: what is it? The question is answered here briefly: logic is nothing more or less than a set of rules for relating symbols or statements, propositions in particular; and those rules are limited only by the human imagination. Some critics of the sign rule have rejected that definition, but they do not state the rationale, much less an alternative definition. The stance of the critics is all the more puzzling, given their inclination to suggest that mathematics provides a mode of theory construction suitable for sociology. The immediate point is that mathematics qualifies as logic in accordance with the definition of logic just given, but for reasons that need not detain us some philosophers (logicians in particular) and some mathematicians would dispute the definition and/or its application to mathematics.

Returning to the definition of logic, there is only one defensible question about any set of rules for relating symbols or statements: can they be applied *consistently* in the pursuit of a particular goal? As clearly suggested by use of the sign rule in Chapter 6, the goal is to detect *possibly* demonstrable associations between quantitative variables, and critics of the sign rule have refrained from so much as suggesting that the rule cannot be applied consistently. What they are objecting to has never been made clear, but some possibilities are examined subsequently.

Preliminary to examination, consider just one of a multitude of relevant questions: exactly how could mathematics be used to state the illustrative theory in Chapter 6? Reconsider Postulate 1: Among countries, the greater the class conflict in a country, the greater the income inequality in that country. Now if "class conflict" is "*C*" and "income inequality" is "*I*," you can surely imagine an equation something like this: I = 7.35 + 2.43C. But what happened to the unit term (i.e., "Among countries")? Even if we allow proponents of the use of mathematics in stating sociological theories to say "Never mind," what could possibly be the provenance of "7.35"and the "2.43"? Anything like such numbers would have to emerge from research on the empirical association *among countries* between measures of class conflict and income inequality. But there are no measures of "class conflict," and even if there were it is surely difficult to believe that the research would yield the kind of numbers required for the equation in question.

So, believe us, equations fit quantitative sociological generalizations like a saddle fits a cow. As for statements in the form of "*Y* is a function of *X*" or "*Y* is an inverse function of *X*," they are no less objectionable than are the prescribed relational terms (i.e., "greater ...

greater" and "greater ... less"). Indeed, the meaning of the "function expression" and the prescribed relational terms is exactly that of a third alternative: "*X* varies directly with *Y*" and "*X* varies inversely with *Y*."

An Alternative No Less Grotesque than Mathematics

Of the critics of the sign rule, only H. Blalock (for references to his relevant publications, see Alessio [1982]) has seriously pursued an alternative—the use of a causal language. Unfortunately, Blalock did not appear to be fully aware of the seemingly irresolvable issues and insoluble evidentiary problems that have haunted the notion of causation ever since David Hume's treatment of the subject. Most important for present purposes, at no point in his publications did Blalock recognize that there is nothing approaching a genuine "logic of causal relations," unless one confuses it with the "statistics of causal relations" (e.g., as in Pearl 2009).

More need be said about the last point. Consider two assertions in this form: *X* causes *Y* and *Y* causes *Z*. Surely both assertions ignore various distinctions, such as that between necessary and sufficient causation. Never mind that the distinctions become complex, difficult, and debatable when applied to the relation between *quantitative* variables, especially when it comes to deductions from causal premises. Just assume necessary *and* sufficient causation has been asserted. The question then becomes: given the corrected version of the assertion to read "*X* is the necessary and sufficient cause of *Y*" and "*Y* is the necessary and sufficient cause of *Z*," by what rules can one deduce "*X* is the necessary and sufficient cause of *Y*"? Of course, we could invoke "indirect causation," but in the way of a strict logic we will have cracked open a bucket of worms. The immediate problem is that the term "causal relation" does not denote a kind of logical relation, and it can be made to do so only by fiat. And mind you we have not even considered defining causation (including various types of it, such as "partial") and all of the problems in formulating evidentiary criteria. The contention is not a denial of the possibility of a truly complete and defensible "logic of causation" as applied to quantitative variables. Rather, Pearl (2009) notwithstanding, such a logic is simply not out there, and good luck in trying to formulate one.

Bottom Line on Logic

Why belabor the absence of alternatives to the sign rule? Because "alternatives" is exactly the subject that the vast majority of critics of the sign rule ignore. Why the indifference to the subject? Because the critics ostensibly have little time for the very notion of logic. Consequently, they fail to see not only that the sign rule is a form of logic but also see no need to replace it with some other form if it is defective. That the replacement must be appropriate for stating sociological theories is only one of a multitude of problems.

So why advocate use of the sign rule in stating quantitative sociological theories? Because, like it or not, that is the best sociologists can do, perhaps for generations. However, that is not to deny that there are complexities and problems in the use of the sign rule, a subject to which we now turn.

Complexities, Issues, and Problems about the Sign Rule

To make the point again, from sociology's very beginning virtually all theories have been stated in accordance with the conventions of a natural language. Given that tradition, any proposed mode of formal theory construction (MFTC) is likely to encounter stiff resistance. However, some resistance will stem from objections to specific components, as opposed to merely a defense of tradition. Of those components, one very likely objection will be to the prescribed use of the "sign rule."

Complexities that Warrant Special Recognition

To repeat from Chapter 6, as regards application of the sign rule, when the relational term of a premise is "greater … greater" the premise is assigned a plus sign (+) and when "greater … less" a negative sign (−). The rule applies when and only when there are two or more premises linked through a common constituent "unit term" and at least one "characteristic term," be the latter a construct, concept, or referential. So there is only one application rule that generates a conclusion (designated as a theorem) from such a series of linked premises. Stated as simply as possible but more elaborately than in Chapter 6: the conclusion deduced from a series of linked premises has a negative sign when and only when there is an odd number of negative signs in those premises; otherwise, when all of the signs in the premises are positive *or* there is an even number of negative signs, the sign of the conclusion is positive.

To make the point again, critics of the sign rule never so much as suggest that it cannot be applied with consistent outcomes, meaning agreement in independent applications to premises stated as prescribed in Chapter 6. So what are the objections? Again, they are by no means clear. For example, Alessio (1982: 114) refers to the sign rule as "mysterious," apparently *denying* that there can be forms of logic and accompanying rules *other than* those unquestionably conventional (certainly the classical syllogism and mathematics). If so, some of the other critics join Alessio in that denial (especially H. Blalock; see references in Alessio 1982).

The fundamental consideration in this debate is that critics of the sign rule simply do not see that it is used *for nothing more* than detection of possible empirical relations between variables. Why? There is only one answer. The critics evidently think of relations asserted in the premises as being statistical associations. The fundamental mistake in that line of thinking is what it ignores. When some if not all of the variables in the premises are neither observable nor measurable, the relations among those variables are *not* like statistical associations, because the latter are subject to measurement and the very idea of demonstrable measurement error does not apply to premises.

The issue can be further clarified by relating the last paragraph to the illustrative theory in Chapter 6. "Class conflict" is not observable, let alone measurable. Otherwise, there would be the distinct possibility of what approaches a direct test of all three postulates in the theory. If that were the case, why would we become entangled in all of the issues and problems in deduction? Forsooth, why would there be anything like an acute need for formal theory construction? The critics of the sign rule write as though the questions never occurred to them.

The mistake of critics of the sign rule leads them implicitly or explicitly to this question: what kind of statistical association among the variables in the premises would justify the deduction of some particular kind of statistical association, such as that expressed by the Pearsonian product-moment coefficient of correlation? The crippling problem is that a defensible answer would require pages and enormous talent in statistics. Why? Because the briefest intelligible answer would pertain to the kind or kinds of relations asserted by quantitative premises, including (1) direction, positive vs. negative; (2) form, nonmonotonic vs. monotonic and linear vs. exponential in the monotonic case; (3) degree of association in each relation asserted in the premises, granted no measurement error; (4) the type of statistic employed to express the association in tests of the conclusions (e.g., some kind of ordinal coefficient); (5) what would be the extent of correlation of error terms from one premise to the next if the premises were testable; (6) the amount of measurement error in the values that entered into *tests* of the conclusions; and (7) the number of premises. In light of such contingencies, it is surely difficult to describe what would and what would not justify use of the sign rule to deduce theorems from untestable premises *other than the sign rule itself.*

The Foremost but Concealed Issue

If the justification of the sign rule is as simple as just suggested, why are its critics so fixed in their opposition? It is not really the mode of deduction itself; rather, it is something that the opponents of the sign rule do not express and may not be fully aware of themselves. It is what those who employ the sign rule are prone to suggest when outcomes of tests of the conclusions (theorems) indicate substantial predictive accuracy. Unfortunately, the "sign rulers" are tempted to make statements indicating that such test outcomes verify or corroborate the postulates. That is the fallacy of affirming the consequent, meaning the failure to recognize that a conclusion from some set of premises—whatever they may be—could be deduced from *entirely different* premises. Observe that the fallacy has nothing whatever to do with the dilemma of induction, meaning that we have no truly confident basis for anticipating what the next tests of a theory (be they positive or negative) will reveal.

Perhaps critics of the sign rule do not speak of the affirmation fallacy because they recognize that it is literally impossible to demonstrate that some particular conclusion can be deduced from the premises in question and *only* from them. If so, the critics should also recognize that the affirmative fallacy can be avoided only by abandoning deduction as a scientific tool or by especially cautious interpretations of outcomes of tests pertaining to theorems. The more general point is the necessity of recognizing two things. First, the very nature of the content of sociological theories makes deductions imperative for tests of them. Second, until the critics identify defensible alternative to the sign rule, they should learn to live with it. Of course, if they are not really interested in "deductive theories," that is another matter.

Still another concealed issue is that critics of the sign rule never mention the significance of absolutely *negative* outcomes (no predictive accuracy whatever) in tests of a generalization deduced by the sign rule. The extreme negative case would be a descriptive statement in which the direction of association is *opposite* to that predicted by the hypothesis. Such an outcome would clearly indicate that *at least one of the premises* asserts some empirical relation that, if the premise were testable, it simply would not hold. Surely that possibility does not translate as rejection of the falsifiability principle in science, though critics of the sign rule appear oblivious to the point that use of it in stating a theory can result in overwhelmingly negative evidence against the theory. If the sign rule is logically defective, how could its use possibly result in rejection of the premises? Critics of the sign rule cannot attribute negative outcome in a test to measurement error, for that alone would suggest that the sign rule is not *logically* defective. It would be like a blind-from-birth drunk attributing her/his car wreck to intoxication. So it is that the sign rule becomes "defective logic" only when tests support the conclusions reached by its application. Curious is hardly the word for it.

Lest there be some misunderstanding, the foregoing "concealed issues" are very important, but they are secondary to a previous contention, that critics of the sign rule fail to give sociology a feasible alternative. Elaboration requires some repetition, but the importance justifies it.

The use of a causal language as the relational terms of premises makes them untestable even if it becomes possible to measure the constituent variables, at least until there is effective consensus as to realistic criteria of casual evidence, something awaited for centuries. Critics of the sign rule appear indifferent to the problems that are created when a causal language is used in formulating a theory. That indifference is all the more puzzling if one assumes (realistically) that a causal language is one of only two alternatives to something like the language prescribed in Chapter 6, the other alternative being the language of mathematics.

As for stating the premises of quantitative sociological theories in the form of equations, there are three questions. First, how are such statements possible when the vast majority of major sociological terms (e.g., social integration) are not even measurable? Second, without a clearly arbitrary definition of "logic" or a feckless appeal to convention, how can deductions from equations be "logical" but not those from application of the sign rule? Third, how can the language of mathematics, equations in particular, be used to express a causal relation? Yes, the questions are rhetorical in that readers will likely know the answers.

Limits to the Utility of the Sign Rule

Although the sign rule is useful in deducing *ordinal* associations, it has some definite limits. Those limits warrant recognition, but they are a far cry from the usual objections.

Nonmonotonic Relations

The most conspicuous limit of the sign rule is that it can be applied correctly in deducing theorems only when the relations asserted by the premises are *assumed* to be monotonic. However, the difference between a linear and nonlinear association—both being monotonic by definition—is one of degree; and the more linear the relations in the premises, the more likely that in a test of the theory the hypotheses and descriptive statements will agree as to the direction of the deduced associations (those asserted by the testable theorems). That contention stands even though the direction of the relations asserted in the premises cannot be known in any direct sense, let alone with certainty.

Lest there be any confusion, two clarifications are needed. First, the word "association" is used and has been used when the relation in question is presumably measurable, and "relation" when measurement is either not feasible or irrelevant. Second, the relation between two variables is monotonic if and only if the direction of the relation (positive or negative) is more or less the same throughout the *observed* range of the values of both variables. "Observed" is emphasized to indicate that *possible* but unobserved values (again, both variables) are simply irrelevant. To illustrate, whatever the territorial entity and assuming no dwelling places with multiple levels, it is virtually certain that there is an *absolute* maximum for the residential population density (RPD) if only because there are so many feet in a square mile. Yet it is scarcely less a virtual certainty that no actual population has or ever will even approach that maximum. Nevertheless, it is entirely *possible* that for at least some kinds of territorial entities (e.g., census tracts or cities) beyond some RPD level there is a change in *direction* of the association between RPD referents and the referents of some other variable (e.g., age-adjusted mortality rates). Extensive and convincing examples cannot be given because long ago social scientists apparently ceased examining scatter diagrams as a *conventional* activity, even though the examinations would reveal the extent that particular statistical assumptions have or have not been satisfied. Such examinations are virtually certain to become even rarer with emerging popularity of multivariate analyses of Big Data. Such are the difficulties of "doing science."

Final Considerations as Regards the Sign Rule

The real epistemological problem surfaces when in a test of a theory the descriptive statements completely contradict the corresponding hypotheses, meaning that they differ even as to the direction—positive or negative—of the association. There would be no way of knowing whether the contradiction stems from unsatisfied statistical assumptions (the relations asserted in the premises are monotonic and *sufficiently* linear) or simply that the premises are false and perhaps extremely so (i.e., were it possible to measure the variables in the premises, even the direction of the *actual* associations would contradict the premises).

A particular response to the foregoing is anticipated: in using the sign rule by your own admission you do not know the extent that the "relational assumptions" are met by the theory's premises. The reply: that is precisely the reason for the attempt to assess the premises through tests of the implied conclusions (theorems). After all, applications of the sign rule can produce evidence of possible empirical associations that might well have gone undetected. As for the assumption in using the sign rule of *sufficiently* linear relations among the variables in the premises, positive test outcomes do not and cannot verify that assumption. Nevertheless, a test outcome may be such that it contradicts some if not all of the premises, even though the "which one" question is not answerable with any certainty. But are critics of the sign rule demanding the impossible—a demonstration that the conclusions (theorems) can be deduced from one and only one particular set of premises? Finally, granted admitted limitations of the sign rule, if you know of a more appropriate rule for deducing theorems in sociological theories, please inform us; but spare us any proposed use of equations or a causal language in stating premises.

Three Fears about Formal Theory Construction

If and when sociologists contemplate adoption of some MFTC, we anticipate three fears. First, sociological theories would become less plausible. Second, sociological theories would somehow become too complicated, less informative, and boring (perhaps the real dread). Third, the proposed mode of theory construction would be an attempt to impose orthodoxy on the field.

The First Fear

An examination of reactions to the typical sociological theory will reveal abundant reliance on plausibility, much less on testability. Actual systematic tests of a theory (henceforth "sociological" is understood) are rare; hence, its predictive accuracy is little more than conjecture. Instead, theorizing by sociologists appeals to loosely specified and plausible features of human nature or social life. Consequently, "tests" are all too commonly something like: "That seems right." or "Rings true to me."

Such assessments have an empirical character, but they are commonly anecdotal and frequently appeal to plausibility as determined by some overarching perspective in sociology—conflict, functionalism, Marxism, feminism, etc. If the theorist is writing an appeal to the "left," her/his arguments in support of the proposed theory are likely to be anchored in the importance of conflict in human affairs. By contrast, as witness the once popular Parsonian sociology, the arguments of a theorist with conservative leanings are likely to appeal to the consensual quality of social life. Still other perspectives may center on an abhorrence of socialism and appeal to Ayn Rand's rational individualism. And so it goes, providing grist for the mill of the "social constructionists."

Our promotion of formal theory construction truly favors no particular substantive sociological perspective (e.g., functionalism, Marxism, feminism); rather, it stems from rejection of the very idea of assessing theories, whatever the perspective, in terms of their plausibility. Of course, plausibility will enter into assessments of any theory, however stated. Nevertheless, comparisons of contending theories are much more compelling when the contenders are stated formally and the comparison is focused on their predictive implications, test outcomes especially.

In many ways, of the three fears concerning some MFTC, the fear of abandoning plausibility is the least compelling. It is always possible to state a theory two ways, the first in conventional form, building a plausible case for the theory and then translating it in accordance with some MFTC or some mix of the two. The latter strategy was followed somewhat in stating the illustrative theory (Chapter 6), and it warrants consideration all the more because it can be extended to a formal restatement of one of the "grand theories" in sociology's history (e.g., Gibbs [2003] on Durkheim's theory about the division of labor). By preserving some of sociology's past, such extensions would lessen the radical consequences of a turn to formal theory construction, the awful fear of orthodoxy in particular.

The Second Fear

We can hear the critics now: The proposed mode will generate theories that are too complicated, uninformative, and boring. That criticism is debatable, and it certainly ignores the challenging task of comparing theories in light of all dimensions of predictive power (Chapter 4) and the possibility of considering two versions of each theory (*supra*). For that matter, comparing contending modes of formal theory construction is hardly pedestrian, all the more given its novelty.

There is no contradiction in promoting the idea of two versions of a theory, one conventional and the other formal; actually, there are numerous reasons to do so. Even if the formal version is taken as the ultimate goal, the conventional version could stimulate interest in the theory, which leads to its formal restatement and tests of it. Additionally, and even more important, the conventional version might stimulate ideas, some of which could lead to a revision or expansion of the formal version.

If formal theory construction, whatever the particular mode, does entail some loss for sociology, the very least would be rhetorical comparisons of contending theories. The most defensible comparison is possible only when the contending theories have been stated in accordance with the *same* formal mode and, consequently, the comparison can and should consider all dimensions of predictive power.

The Third Fear

Having been exposed to hermeneutics, deconstruction, social constructionism, and postmodernism over several decades, many a sociologist has developed an aversion to orthodoxy. This aversion is expressed in various ways, with railing against positivism being at the forefront.[1] Advocating "ontological and epistemological pluralism," as Abend (2008) does, is another expression of the aversion to orthodoxy. After reviewing seven meanings of "theory" in sociology, Abend (2008: 196) contends that he does not advocate strongly for one approach or another. Instead, he suggests we should "make *as few* ontological and epistemological demands *as possible*." While this leaves theorizing almost wide open, even Abend draws the line at such things as a witchcraft perspective.

Wherever the line is drawn, objections by sociologists to FTC cannot be justified by appealing to pluralism. After all, the currently dominant mode of theory construction is *not* formal; far from it. So developing a niche for FTC would increase sociology's pluralism.

Science: Why Bother?

Whatever the merits of this or that mode of theory construction, formal or otherwise, the question is frequently asked: what is the purpose of scientific theories or, for that matter, science? Purists appeal to an expansion of knowledge. By contrast, political progressives advocate changing the world, while conservatives would opt for the status quo, and reactionaries yearn for a world long since gone, as when U.S. right wing radicals let it slip that their goal is nothing less than repeal of the New Deal. At the risk of startling readers, let it be known that the authors are among the purists.

We know of no recent survey of opinions of scientists that speaks to the "purpose" question, but our hunch is that most scientists would favor knowledge expansion. Consensus on that point would lead to another but related question: Granted that science expands knowledge, to what end? In some instances expanded knowledge, carefully applied, has improved the quality of life by virtually any standard. Early detection and treatment of cancer and satellite-based warnings of dangerous weather come to mind, as does the longer-range and politically charged assessments and projections of climate change.

Should it be objected that there is no effective consensus among scientists on anything, try to find an astronomer who rejects the heliocentric theory. Indeed, the most remarkable feature of science is that its adversarial character has not precluded *eventual* effective consensus on many things, albeit frequently and perhaps fortunately not "total" consensus. Philosophers of science are prone to ignore that point because of their preoccupation with the "problems in science," as though such problems preclude achievements. Say what you will, no matter the answers given to the purpose-of-science question, assessing "purpose" involves value judgments. This being the case, determining purpose is, strictly speaking, outside the purview of science, no matter the effective consensus among scientists. It is more a matter of perspective and political dispositions. But, again, who is prepared to deny achievements in science, especially in light of all dimensions of predictive power (Chapter 4)?

Despite all this, there are numerous contending perspectives in sociology, some of which (1) define the field's subject matter, (2) describe the appropriate kinds of questions, (3) identify the procedures for pursuing answers, and/or (4) identify criteria for acceptance/rejection of an answer.[2] One question posed by the "Parsonian perspective" is something like: what is the basis of social order? The sociological literature frequently suggests two opposing answers: conflict or consensus. So it is that contending perspectives take sociologists in quite different directions, not only as to acceptable questions but also acceptable answers. But the argument here is not against this or that perspective in sociology. We do argue, however, that the merits of a perspective should be judged in light of the predictive power of the theories it ostensibly generates and *not* the reverse!

The foregoing leads to recognizing a link between perspectives and formal theory construction in the way of two contentions. First, if a perspective is to be judged by the predictive power of the theories it generates, those theories and those of contending perspectives must be systematically testable if only in recognition that judgments of predictive accuracy require tests. Second, at least in sociology, formal theory construction is necessary for *systematically* testable theories.

In this regard, the illustrative theory in Chapter 6 takes on added significance because it reflects the conflict perspective in sociology but offers more in the way of testability than do other "conflict theories." Accordingly, assuming that systematic tests of the illustrative theory are undertaken, regardless of what the tests indicate as regards predictive accuracy, the tests alone would point to an advantage of the conflict perspective over the consensus perspective. It is difficult to imagine a formal statement of a theory that reflects the consensus perspective, especially a theory in which "country" or "countries" is the unit term; but it is no less difficult (Collins 2012 notwithstanding) to imagine a *systematically testable* conflict theory in the conventional form (i.e., not formally stated) and at the international level.

Some Inconspicuous Implications

The foregoing brings us to some inconspicuous implications of our MFTC and the illustrative theory in Chapter 6. Space limitations permit a brief treatment of only a few such implications.

The Core Purpose of Science

Would even very positive outcomes of tests of the illustrative theory in Chapter 6 demonstrate that class conflict is even a cause of international variation in income inequality, occupational mobility, and the property crime rate? No! Does the theory otherwise explain the variation in any of those three variables? Again, no!

In light of these negative answers one may wonder, why bother? How could the theory be taken as anything more than as an illustration of a particular MFTC, devoid of any utility whatsoever? The answer is simple: depending on test outcomes, the illustrative theory *possibly* furthers the basic goal of science, which is nothing more or less than the detection of order in events or things and furtherance of that order through novel conceptualizations. As for causation and explanation, they may be eventual possibilities; but as for demanding defensible sociological theories that assert causation or purport to explain, that demand is not likely to be met for generations to come. Moreover, to use a football analogy, in aspiring to score a touchdown, sociologists ignore the more immediate need to make first downs; and ignoring that immediate need is the path to persistent failures.

Substantive Problems neither Avoided nor Solved

It hardly need be said that in formulating a scientific theory, whatever the subject, sociologists confront a sea of problems. As a case in point, no sociologist is likely to be unaware of the "selectivity problem" when analyzing some empirical association, often the first step toward a theory. To illustrate, anyone who has worked in a jail or prison quickly comes to recognize that an exceedingly disproportionate percentage of the prisoners are poor by any standard. That association surely suggests that poverty is a cause of criminal behavior, perhaps even a major cause. But that suggestion ignores selectivity in the process of criminal conviction and sentencing.

There are actually several selectivity mechanisms that could account for at least some of the poverty/incarceration association and *possibly* departures from it. The most obvious is that following an arrest poor suspects are the least likely to mount an effective legal defense prior to and during a trial, for it can be very costly in the vast majority if not all criminal justice systems. For a depressing but convincing demonstration, read accounts of O.J. Simpson's murder trial (e.g., Gibbs 2008: 161–178).

The point is not just that formal theory construction provides no solution to the selectivity problem in contemplating the poverty/incarceration association as a possible component in a theory about crime. Space limitations prevent even their identifications, but all manner of kinds of selectivity problems are neither solved nor avoided by formal theory construction. Nevertheless, the immediate mitigation should be obvious. Those evidential problems are neither solved nor avoided by the conventional mode of theory construction.

More on Epistemological Implications

When considering the illustrative theory in Chapter 6, although far from obvious, there are four links or steps, including epistemic statements, between "class conflict" and any referents that *would be* computed to test the theory. That awful "distance" between abstract notions and referents is very likely to obtain in test of any formally stated sociological theory. Hence, one must rightly wonder as to how any measure of association between sociological variables can even approach the maximum (usually +1.0 or −1.0). And there are so few reports of such an association that the audience is likely to suspect that any instance is spurious or reflects a nest of tautologies.

Frequently cited reasons for the usual meager associations among sociological variables are "failure to control for exogenous variables" and/or "measurement error." Such possible reasons surely cannot be dismissed, but there is a seldom recognized and even more disturbing argument: The meager associations reflect the extremely abstract quality of major sociological constructs ("class conflict" being only one of many) and their questionable empirical applicability. Herein may lie the continuing commitment of sociologists to the conventional form of stating theories; it mercifully conceals the enormous problem of empirical applicability and the need for some systematic way to link constructs with empirical referents.

Until something else comes along, the clearest path to such construct/referent linkage is formal theory construction, and if adopted it will not have created sociology's bevy of extremely abstract notions. They have been firmly entrenched in the field's vocabulary from the very beginning. Even so, sociologists are unlikely to adopt any MFTC until they take predictive power seriously, especially the testability dimension.

Multivariate Analysis Replaces Sociological Theories

Adoption of formal theory construction (FTC) is likely to be perceived by sociologists as a rejection of "grand theories," and that is the case largely because the grand theories were stated in accordance with the conventions of a natural language. But over the past half century another tradition has appeared in sociology, one that stands in opposition to both FTC and grand theory. That rival is "multivariate analysis," but the rivalry itself has gone virtually unrecognized in the sociological literature.

A Conspicuous Tradition

Although commonly unrecognized, multivariate analysis (MA) has become a definite tradition in sociology if only in being so distinctive.[3] In its purest form it is the use of regression coefficients to describe the association between each of two or more "independent" variables and another variable identified as "dependent." Although not a distortion, that characterization ignores the belief that, depending on its regression coefficient, each independent variable is at least *a* cause of variation in the dependent variable. If so, the belief reflects not only uncritical thinking about causation but also the absence of even effective consensus in all of the sciences as to criteria of casual evidence. The comment extends also to applications of linear models, logistic regression, stepwise regression, or survival analysis (Harrell 2001).

The issue in the case of MA by sociologists is definitely *not* the identification of the variables as "sociological." Of course, that identification is debatable because it relates to a question: What is sociology's subject matter? It would be rash to deny the importance of the question, as the coherence of the field (or lack of) depends largely if not entirely on the answer. Nevertheless, grand theories, MA, or FTC do not offer an answer to the subject matter question, at least not one that the majority of sociologists are likely to accept. It is simply a question that haunts the field, and the abiding fear of sociologists of orthodoxy does not encourage attempts at an answer.

What has been said of sociology's subject matter extends to issues/problems inherent in the notion of explanation, not only as to its nature but also criteria of an adequate explanation. All too briefly, it is simply wishful thinking to argue that MA solves any of those problems or answers any of the pivotal questions (e.g., for an explanation of an event or thing to be adequate, must it assert necessary *and* sufficient causation?).

The Most Immediate Issue

In using regression equations sociologists have borrowed indirectly from mathematics and directly from statistics. The latter is the most strategic because statisticians have identified the crucial assumptions.[4]

The pivotal assumption pertains to a possible interpretation of a regression coefficient. It expresses some feature of the association between variables among entities, necessarily a particular finite set of them. If interpreted as indicative of the association between the same variables in some larger set, the interpretation is justified only insofar as the particular set is a random sample of the larger one.

Needless to say, "sampling" is a very complex technical subject. Nevertheless, it can be said with confidence that the "random sample" assumption is violated more than any other made by sociologists, along with other social scientists, even econometricians. It is as though they say "Well, the entities in question are not a random sample, but let us pretend otherwise and compute the test of significance for the association anyway."

The suggestion is not that the problems entailed in MA (the "statistical" is understood) would be solved if sociologists analyzed data only for entities that are truly random samples. Sociologists commonly appear indifferent to the point that sampling and tests of significance are first and foremost devices to reduce research expenditures. So why invest the resources required for, say, a survey of all residents of New York City when reasonable inferences can be reached by a survey of a random sample of those residents? The inference becomes dubious when extended beyond New York City or when the sample was not selected on a truly random basis.

Other Technical Problems Pertaining to Multivariate Analysis

Some sociologists who engage in multivariate analysis (MA) do not appear fully aware that interpretations of regression coefficients, multiple correlation coefficients, or partial correlation coefficients entail more assumptions than "random samples." A very important additional assumption is that the values of the variables are normally distributed. Only rarely does a MA report express awareness of that additional assumption, and it is even rarer for a report to provide supporting evidence. What has been said of the "normal distribution assumption" extends to another crucial assumption, that the associations in question are linear.

Still another but perhaps less recognized assumption is that the association of two variables, *X* and *Y*, is more or less the same at all values of *X* and *Y*. Even far less recognized, there is the assumption in conventional statistics that one variable, *W*, cannot influence the association between *X* and *Y* unless *W* is associated with both variables. That assumption is avoided by the "set comparison method" (SCM), a distinct alternative to conventional MA.

When employing the SCM, the investigator creates sets of units such as to maximize *interset* contrasts as regards either the *magnitude of or the variance in* some variable, *W*. He/she then examines the association *among the sets* (i.e., interset) between some intraset attribute of *W* (mean or standard deviation) and the intraset association of two other variables, *X* and *Y*. The set attribute of *W* may vary directly or inversely with the amount of intraset association between *X* and *Y*, even though there is no association whatever between *W* and either *X* or *Y*.

A "real world illustration" comes (Gibbs 1989: 158) from the creation of four sets of countries, such that Set I comprises the 26 with the greatest per capita international trade (assumed to reflect the *least* economic autonomy) and ending with Set IV, the 26 with the least trade. For each set *rho* was computed for two variables: (1) per capita consumption of inanimate energy (the energy variable, postulated as reflecting the efficacy of inanimate control) and (2) the percent of the economically active in agriculture, forestry, fishing, or hunting (the biotic variable, postulated as reflecting the *inefficacy* of biotic control). The *intraset* correlation (again, *rho*) between the energy variable and the biotic variable varies from −.433 (Set I) to −.809 (Set IV). The *interset* correlation (*rho*, N=4) between the mean trade value for a set and the energy/biotic *rho* for the set is −1.00, evidence that in an ideal condition—absolute economic autonomy—the *negative* association among countries between the energy variable and the biotic variable would be very near if not at the maximum.

Unlike conventional MA, the SCM offers an opportunity for an approximation of a genuine experiment in what is an observational science, for it entails actual manipulation of the values of variables. That manipulation also makes it possible to examine an association (here, between *X* and *Y*) in an approximation of an ideal condition (some maximum or minimum *intraset* attribute of *W*). Needless to say, a far from maximum association between sociological variables may obtain precisely because the context is not an ideal condition, and the SCM makes it possible to examine the association in varying approximations of some ideal condition (e.g., a country wherein there is no stratification).

The Necessary/Sufficient Distinction

A conceptualization of causation should recognize the distinction, but it is lost in conventional statistical measures of association. The loss has been especially promoted by MA, for it has reduced the use of scatter diagrams, the very thing that may throw light on the necessary/sufficient distinction even though the variables are quantitative. Only *positive* bivariate associations need be considered.

Think of a scatter diagram in which, as convention has it, the *Y* values are on the vertical axis and the *X* values on the horizontal. If a relatively high *X* value is necessary but *not* sufficient for a relatively high *Y* value, then the entities (whatever they may be) will be largely if not entirely *below* the diagram's geometric diagonal, a line connecting the lower-left corner of the rectangle with the upper-right corner. Stated another way, as the *X* values increase, they become more necessary but less sufficient for increases in the *Y* values.

Conversely, if a relatively high X value is sufficient but *not* necessary for a relatively high Y value, the entities will be largely if not entirely *above* the geometric diagonal. Again stated another way, as the X values increase they become more sufficient but less necessary for increases in the Y values.

When analyzing quantitative associations, sociologists are not prone to recognize the necessary/sufficient distinction as manifested in a scatter diagram. Specifically, the necessary but not sufficient condition will be manifested as an exponential association, while the sufficient but not necessary condition will be manifested as a logistic association.

In both cases, the logistic and exponential, the product-moment coefficient of correlation (r) will be less than maximum because the X/Y association is monotonic but nonlinear. Nonetheless, the distributions of the values in a scatter diagram would indicate that the necessary/sufficient distinction is relevant, and the *ratio* of the number of entities above the diagram's geometric diagonal to the number below would make the distinction a matter of degree. Whatever that ratio, given that only two variables (X and Y) are involved, MA would be irrelevant.

The necessary/sufficient distinction offers a basis for deducing a logistic or exponential association, the mark of a sophisticated theory. Rather than recognize that point, sociologists are puzzled by a logistic or exponential association, and they are prone to make instances linear by some technique (e.g., logging values). So it is that technique/method trumps substance.

The necessary/sufficient distinction takes on great importance in "doing science," a point seemingly lost on those who emphasize "genetic determinants" of human behavior. Yes, in a limited sense virtually all human behavior is genetically determined, and doubts can be reduced by attempting to teach a chimpanzee how to play poker. But even casual observations on a variety of human populations will suffice to indicate that any particular genetic composition is at most only a *necessary* condition for playing poker.

The Desert Created by Conventional Multivariate Analysis

Even if all of the problems entailed in MA could be justifiably ignored, the most severe criticism would stand. Stating that criticism as briefly as possible: MA has turned sociology into a desert as far as theories go. Defenders of MA will claim that, unlike the grand theorists, the practitioners are "doing real science," but a theoretically sterile science is itself sterile.

Should the criticism be doubted, observe first of all that there is no "unit term" in regression equations. Yes, the report of a multivariate association is likely to reveal that the values of the variables are attributes of some particular kind of entity, the "unit term" in the present parlance of FTC; but the report probably ignores this question: why that kind of entity rather than another? All too frequently the answer has to do with data availability, the trademark of dustbowl empiricism. Above all, it is a virtual certainty that the report does not assert "greater association of variables among the designated kind of entities than among alternatives." Bear in mind that such *specificity* is a dimension of predictive power, and there is no more compelling rationale for the choice of a particular unit term in stating a generalization or theory.

Still another question reveals a major problem with MA. Whatever the entities, think of a regression equation in which "education level" and "income level" are the independent variables and "frequency of voting Republican" is the dependent variable. Why expect any association? An answer is virtually certain to introduce a fourth variable, one that is assumed to link both independent variables with the dependent. If those links are couched in a natural language, they would be less intelligible than a rendition in the language of FTC. The illustrative theory in Chapter 6 notwithstanding, that language does not preclude premises that assert something more than bivariate associations.

The foregoing should not be construed as anything like a categorical dismissal of conventional statistics. To the contrary, should scientists adopt relative predictive power as the supreme criterion for assessing generalization or theories, there would be no reason for abandoning conventional statistical terms (e.g., correlation, mean, mode, standard deviation) when undertaking tests. Actually, there is no realistic alternative.

As for MA, there is no denial that it may be useful for all sciences, the immediate reason being that a theorist may stipulate the use of some form of MA to test her/his theory. Of greater importance, a particular MA may point to uniformities that pose a challenge for theorists. Viewed in that light, MA would not have created a desert for sociological theories had the practitioners recognized that their findings were only steps toward theories.

Penultimate Parting Words

To the surprise of no one, we have treated "additional issues" with a view to promoting acceptance *in sociology* of our MFTC. Nevertheless, our fear is that the mode will be construed as somehow set in stone. Nothing could be further from the truth.

Alternative Modes of Formal Theory Construction

Perhaps most important, our ultimate goal is to sell FTC, and to that end we are truly predisposed to consider alternative modes, hopefully one that would dampen objections to FTC itself. One possible objection to our mode is its complexity.

Although we believe that the mode's complexity is less than it may appear, we would surely entertain simpler modes, just as long as they do not preclude achieving what we consider as the two primary goals of FTC (whatever the mode). The first is to promote the testability of sociological theories, and the second is to realize much greater consensus in assessments of sociological theories. The latter goal is the primary reason why we would not abandon the predictive power criterion, whatever the mode of theory construction, but note that the criterion is compatible with all manner of FTC modes. It is even compatible with the conventional natural-language mode but much more difficult to apply in that case.

Of course, prospects for alternative modes of FTC will change as sociology changes. One major change would be a movement toward something that sociology scarcely ever enjoyed, that being a long natural history stage. The movement would promise a vast conceptual clarification and an identification of terms that offer more empirical applicability than does sociology's current terminology. That change would in turn set the scene for a MFTC that does not entail constructs. A mode that recognizes only concepts (as "characteristic terms") would identify all generalizations as propositions (i.e., no axioms or postulates) and for that reason alone it would be considerably simpler than our mode. Unfortunately, the prospects for a natural history stage (NHS) appear extremely remote; but even consideration of formal theories would force recognition of what could be gained by a long NHS, simplification of the extrinsic parts of theories being the first gain.

Other Issues and Problems Reconsidered

The prospects for particularly relevant changes in sociology even include drastic improvements in conceptualizations of causation, including the development of empirically applicable and realistic criteria of causal evidence. Such development would set the scene for deducing testable theorems from *causal* premises and possibly abandoning or revising the sign rule. Steps in that direction have already been taken by one of the authors (Gibbs 1994: 317–328); but even if space were not a real limitation, treatment of the subject in this book would confront the reader with still another and very possibly an intolerable complexity.

Some Limits in Formal Theory Construction

Consideration in this chapter of additional issues is especially incomplete in not recognizing what our mode of theory construction (MTC) will *not* do. Above all, it promises to promote the testability of sociological theories, though not necessarily their predictive accuracy. Even so, in requiring a theorist to formulate instructions for computing referents, our MTC would force theorists to abandon their traditional indifference to empirical applicability, the problem of measurement in particular. As for importance, we freely grant that a truly sweeping solution of sociology's measurement problems would perhaps contribute far more to sociology's progress than would even wholesale acceptance of our MTC.

But consider the possibility of FTC alleviating sociology's measurement problems. Those problems commence with recognition that the field has few genuine measures (especially if demography is not taken as a subfield), a point suggested by the admirable survey conducted by Charles Bonjean, Richard Hill, and Dale McLemore (1967). Rather, what passes for a "measure" in sociology is often really an asserted *correlate* of something designated by a construct or concept, and the common use of the word "indicator" is a semantic dodge rather than a real solution of a measurement problem.

To clarify, there is no denying that "cohesion" is an important notion in sociology, and our MTC forces recognition of the need to identify the most appropriate unit term in connection with cohesion. Why not recognize that it is much more an empirically applicable characteristic of primary groups than of various other entities (e.g., countries or even cities)?

Regardless of the kind of entity, there will be the immediate and seemingly insoluble problem of framing an intelligible, complete, and empirically applicable definition of cohesion. So for that reason alone there is scarcely any prospect of developing a defensible and feasible measurement procedure. Perhaps that recognition leads sociologists to write something like this: In the case of primary groups the frequency of member interaction is an indicator of cohesion.

That statement gives rise to three questions. First, is the statement analytic or synthetic? Second, if analytic, why does sociology need two terms, "cohesion" and "frequency of interaction"? Third, if synthetic, how can evidence be brought to bear on the statement through systematic deductions? Our MTC answers the last question. First, identify "cohesion" as a *construct* in this postulate: among primary groups, the greater the cohesion of a group, the greater the frequency of interaction among the group's members. Second, formulate at least one more "Among primary groups" postulate in which "cohesion" is again a construct but the concept is something other than frequency of interaction, perhaps "membership stability." And, third, deduce a theorem from the two postulates.

Yes, out of fear of appearing pompous, some sociologists might refrain from using a formal mode in the way suggested by the "cohesion case." Even so, appearing pompous is surely preferable to looking like you are functioning with only a room temperature IQ.

On Matters of Greater Importance than Formal Theory Construction

We have admitted that a sweeping solution of sociology's measurement problems would contribute more to the field than would acceptance of any mode of FTC. However, given severely limited prospects of an enormous increase in sociology's research resources, even a substantial reduction in the field's measurement problems is virtually beyond imagination. Yet there are real prospects for what might well be an equally strategic turning point in sociology—launching a full bore search for uniformities in whatever is taken as sociology's subject matter.

We view the search for uniformities as so important and complex that much of Chapter 9 is devoted to it. For the moment we are content with a few general observations beyond the point that an all-out search for uniformities would be a major part of what sociology missed entirely, that being a long natural history stage. As argued previously and all too briefly, from sociology's very beginning theorists occupied central stage; consequently, extensive and systematic searches for uniformities have been very rare. So no one should wonder why several generations of sociologists produced little more than a grab bag of seemingly untestable theories, or why interest in theories has evidently declined in sociology over recent decades. The point is that, whatever the field, the principal function of theories is to explain uniformities by deducing them, which may be the only way to avoid insoluble problems with the notion of explanation.

Notes

1 The strangest feature of the railing is that over past decades several commentators (e.g., Halfpenny 1982) have demonstrated an extreme diversity in the way that authors use the word "positivism," a diversity so great as to suggest that the word is unfit for communication. If there is a common denominator as to the word's meaning, there is only one obvious possibility—antipathy to science.

2 The choice of the word "perspective" over "paradigm" reflects Masterman's demonstration (1970) that Kuhn used the latter word such as to indicate over twenty distinct meanings.

3 For all practical purposes everything said here about multivariate analysis in sociology extends to the social sciences generally. Anthropology is by far the most possible exception, but it has never been known for acceptance of quantitative research methods.

4 The "assumptions problem" is magnified by the sheer number of independent variables that may and commonly do enter into a regression analysis. Indeed, variables can be and often are entered without any particular justification, something that underscores the atheoretical character of multivariate analysis. That character is especially conspicuous when social scientists mine Big Data sets and proceed with algorithms of stepwise regression. However, it is no less conspicuous in several traditional lines of research, such as studies of variation in homicide rates. Tcherni's study (2011) may appear exceptional in that she claims (2011: 475) to have introduced a "multilevel theoretical framework," but it is not stated such that the findings can be deduced from it by any rules of deduction. Nevertheless, it is instructive to read her comment (2011: 476) on a well-known multivariate analysis of homicide rates, that by Land, McCall, and Cohen: "Land and his colleagues intentionally based their research solely on investigating empirical data patterns without imposing a theoretical structure on their analysis."

8

DISASTROUS BELIEFS IN SOCIOLOGY

The "disastrous" in this chapter's title is likely to jar, bringing to mind hurricane Katrina or earthquakes in Nepal. Whatever the image, "disastrous" implies great damage. In this instance the damage done by subsequently described beliefs is to sociology's scientific integrity. If those beliefs continue to flourish, there is no prospect whatever of sociology even approaching scientific standing. Of course, whether sociologists really care about the field's scientific standing is another question.

Belief One: There Is No Objective Reality

A real downer for any science, sociology in particular, comes from the epistemology of solipsism, the belief that denies any reality outside the mind. There is a more moderate version of solipsism, one that grants objective reality but denies human access to it. For all practical purposes, the distinction between the two versions is inconsequential; and the same applies to the belief that the social world is fundamentally different from the physical world (treated rightly as a myth by Heidtman, *et al.* 2000: 3).

One version or another of solipsism is found in Hinduism. Solipsism evolved also among Greek philosophers, and it surfaced spectacularly much later in George Berkeley's celebrated immaterialism. More recently, deconstructionists, postmodernists, and heralds of "social constructionism" have promoted solipsism—wittingly or not—in what has come to be known as the "science wars" (Ross 1996, K. Parsons 2003). Unfortunately, but unsurprisingly, the science critics have not informed their many followers in sociology that they are peddling stale goods, old hats. Lest it appear that solipsism is alien to the sociological literature, consider this remarkable contention: "since there exist great disagreements between interested parties in the categorizations of real-world cases, 'suicides' can generally be said to exist and not to exist at the same time …" (Douglas 1967: 196).

The rationale for denying objective reality is amazingly simple. Whatever we sense of the world passes through our mind, and all minds are shaped by a myriad of forces, personal and collective. Personal bias, cultural shadings, religious beliefs, and/or the influence of power always filter what we know about the world (or think we know). Through these filters, "reality" is socially constructed; so the argument goes. But that argument will not be really understood unless it is recognized as being a vast attempt to make objective reality nothing more than a matter of language (see Wootton's critique [2015] of that attempt, especially the role played by Wittgenstein and many sociologists).

Evidence of a World beyond the Mind

For all but the determinedly obtuse or, paradoxically, the extremely informed skeptic, evidence of objective reality is abundant. Observe, for example, the ravages of the Ebola virus, or with tongue in cheek, invite modern solipsists to step off the twentieth floor balcony to test the social construction of gravity.

The last example is apt because there are forces at work—manifested by order and predictability—whether we think so or not. In a very real sense science is the pursuit of a better understanding of those forces, and its organization minimizes the personal and cultural influences on that understanding.

When doing science, participants design measurements and perform tests in accordance with conventional procedures. These procedures *minimize* the impact of personal preferences, cultural shadings, religious beliefs, and the sometimes corrupting influences of those funding the research. Note the emphasis is on "minimize." Scientific procedures substantially reduce but frequently do not entirely eliminate the "externalities" that influence choices in and outcomes of scientific activities. Hence, science-based knowledge is progressively shaped by outcomes of bias-minimizing procedures. Stated otherwise, the history of science is a history of how scientists have "learned how to learn," as well as gaining a better sense of the limits of that learning (see Weinberg 2015).

Data and Objective Reality

In sociology there is still much to learn about how we collect data on objective reality, and studies are designed to describe the detailed mechanics of a discursive clash of subjectivities (e.g., Uhlmann 2004: 80; Bittman and Pixley 2004). Moreover, resistance to the idea of an objective reality is becoming increasingly questionable in light of an expansion of data available for sociological research and theorizing. Granted, data interpretation is necessarily perceptual and, hence, "constructed," are we to believe that nothing is really "out there" to interpret? One goal of science is to reduce subjectivity by examining whether this or that "constructed interpretation" has predictive implications, with the subsequent goal being an assessment of those implications through general observations or systematic tests.

There is no denying considerable "interpretive noise" in data, but some progress is evident. Over recent decades sociologists have gained access to an enormous body of data that could be used to formulate or test generalizations about the social world. What are we to make of these data? If the world has no accessible objective reality, why do we believe that data are compiled and literally exist *external* to us? Is that belief also merely an illusion?

Particular Bodies of Available Data

The strengths and weaknesses of survey research have long been recognized and refined (e.g., Choi and Pak 2005). Drawing on those recognitions and refinements, since 1972 the General Social Survey (published at the University of Chicago) has been monitoring societal change and studying the increasing complexity of American society, the product being a virtual mountain of publications.[1] The *Social Indicators Research Journal* has been reporting research on the problems related to the measurement of numerous aspects of quality of life since 1974. A series of National Longitudinal Surveys, "designed to gather information at multiple points in time on labor market activities and other significant life events of several groups of men and women," are widely available.[2] Demographers routinely plumb data accessible through such sources as the Demographic and Health Surveys,[3] the American Community Survey,[4] and the Integrated Public Use Microdata Series.[5] Since 1976 the Consortium of Social Science Data Archives (CESSDA) has served as an informal umbrella organization for the European national data archives.[6] The mining of so-called Big Data to map social networks in numerous settings is a promising field in its early infancy (see Baesens 2014). The list goes on and on.

In other arenas of measurement and tests, social scientists are working with neuroscientists in developing techniques to map the neural basis for child development,[7] as well as adult moral and economic decision making (Greene 2014; Moll, *et al.* 2005; Gazzaniga 2005; Camerer, *et al.* 2004). While these measurement techniques and the associated conceptual frameworks are in very early stages, the innovative potential for objectively measuring long elusive social phenomena is promising.

In the realm of ethnographies, researchers have developed the systematic use of multiple data collection methods, both qualitative and quantitative, to minimize personal bias when interpreting life in local communities and sometimes suggesting generalizations (Schensul, *et al.* 1999). Such studies have been collected in the Human Relations Area Files (HRAF) and the Outline of World Cultures, all stimulating scientifically useful research (but recall limitations described in Chapter 4) in light of the following criteria:[8]

Maximum cultural diversity: the cultures should represent, as far as possible, the known range and variety of cultural types in terms of language, history, economy, and social organization.

Maximum geographical dispersal: the cultures should be geographically representative of all major world areas and all major ecological settings.

Adequacy of literature: within the scope of the two preceding criteria, the cultures should have a quantitatively and qualitatively adequate literature coverage.

Special initiatives: from time to time HRAF undertook to build special kinds of collections such as: immigrant and other subcultures within North America, cities, and countries files.

The foregoing treatment of measurement strategies in sociology and cousin disciplines—economics, anthropology, and political science—is all too brief. Any experienced ethnographer, psychologist, political scientist, economist, survey researcher, or demographer could point to a variety of measurement achievements that have not been even mentioned. Nevertheless, we are continuing to learn how to learn, and those critics who ignore or deny greater opportunities to conduct research on important dimensions of social life are simply obtuse.

Finally, even if data are nothing more than "social constructions," it is inconceivable that they are *not* needed to test generalizations or theories; and credulity is strained by the contention that test outcomes are somehow socially/culturally determined. For that matter, unless one assumes some objective reality, it makes little sense to conduct tests or anticipate agreement in assessments of outcomes.

The Numerical Trend in Sociological Theories

Given the growth in accessible data, has it generated an increase in sociological theories over recent years? The question led us to examine items published in the *American Journal of*

Sociology (*AJS*), comparing two time periods: January, 2000, through November, 2004, with January, 2010 through November, 2014. For each period we counted the articles and comments with either or both words "theory" and "theories" in the title/subtitle and calculated a ratio of two numbers, the 2000–2004 number divided by the 2010–2014 number. That ratio is 5.0, indicating a substantial decline of interest by U.S. sociologists in "theory" or "theories" in just one decade. We then did parallel calculations for the title/subtitle of books reviewed in the *AJS* issues. The corresponding ratio was 3.17, again indicating declining interest in theory or theories.

For the same time periods, the manifestation of an ostensible decline of interest in theory or theories commenced somewhat earlier and was far less evident in articles and comments in the *American Sociological Review* (*ASR*). The ratio for the 2000–2004 period to the 2010–2014 period is only 1.20, but it is 2.40 for the 1995–2000 period to the 2010–2014 period.

Obviously, a thorough trend analysis would require a major research project, one that considers all prominent sociology journals, including *Social Forces*, and extending to subfields of the discipline, as well as to other countries. In addition, the research should go beyond titles/subtitles to an in-depth content analysis. After all, the word "theory" or "theories" in a title or subtitle is only one indicator of interest and emphasis.

Both limited space in this book and limited research resources precluded the kind of needed research, all the more because the research should answer another question: what has replaced sociological interest in theories? On the basis of impressions stemming from very limited surveys of *AJS* and *ASR* issues, it appears that in recent years U.S. sociologists have become increasingly interested in gender and race/ethnicity, discrimination in particular. Unless one is indifferent or hostile to democratic principles, those interests are surely commendable; but they have not resulted in a proliferation of related theories.

The paucity of theories on gender and race/ethnicity is regrettable, but there is a positive side to the growing focus in sociology on something. Otherwise, it would be difficult to discern any unity whatever as regards the field's subject matter. To illustrate that point, a diligent search of sociology journals over recent years will find articles on the fate of old typewriters and the first names of babies. (Reference provided on request.).

Our conclusion from this brief excursion: Greater availability of data has not increased interest of sociologists in formulating or testing theories; and increasing rants against "positivism" in recent decades are hardly substitutes. If critics argue that our attempt to promote formal theory construction would terminate interests of sociologists in theories, it is surely difficult to see how the trend could be made worse.

Belief Two: Causation Is Readily Demonstrable

There are numerous sociologists who evidently believe that causation among sociological variables is readily demonstrable. That belief is dangerous in that it encourages indifference to a host of issues and problems concerning causal evidence. Without a substantial reduction in those issues and problems, effective consensus as to criteria of causal evidence cannot be realized; and unless the criteria are realistic, demanding "causal theories" is pointless.

Inherent difficulties in securing causal evidence abound, and any hope of addressing them requires a focus on space-time relations (STRs) as possible manifestations of causation. There are those who *deny* (contra Hume) that causation is manifested *in and only in* STRs, and that belief precludes effective consensus as to causal evidence. By contrast, as Goldthorpe (2001: 2) has noted, "causation must in some way or other imply association." We need to consider some details in connection with the issue.

Mechanistic versus Stochastic Outcomes

It is tempting to assume that if an event, condition, or particular set of circumstances "causes" some identified effect, then the effect will invariably follow. Mechanistic billiard ball analogies come to mind. This rarely happens in settings of interest to sociologists. Instead of mechanistic inevitability, STRs reflect stochastic processes, and predictions must be couched only in terms of increased probabilities that some particular outcome will follow given some assumed cause.

Evidence of the increased probability of an outcome, given the presence or past occurrence of an asserted cause, is just the beginning. While causation must in some way imply some kind of association, an ancient bromide almost automatically follows—correlation is not causation. That bromide ignores Goldthorpe's point (*supra*); and it surely does not speak to the pivotal question: What is evidence of causation?

If you tire of serious discussions of causation vs. correlation and need a break, you might click your way through Tyler Vigen's *Spurious Correlations* as presented on his website.[9] There you will find the astonishing correlation between the divorce rate in Maine and the per capita consumption of margarine for the years 2000–2009, with r =.99. You will also find, during this same time period, that the per capita consumption of mozzarella cheese and the number of civil engineering degrees awarded were highly correlated (r =.96). These findings are interesting and fun, but they certainly do not speak to this question: how do we somehow know that the associations in questions could not possibly have been causal in any sense? The only truly defensible response: well, we do not really know and will not know until we pose and answer another question: do the correlations obtain in other places and times? That is the crucial question, and bromides about correlations and causation provide no answer.

Specifying Space-Time Relations and Type of Entities

To secure evidence of causation we need to speak clearly about the STRs between or among the variables in question, as well as the kinds of entities involved (e.g., countries, cities, organizations, individuals). The argument is best expressed as this book's principal causal contention (PCC): *For any pair of variables the possibility of causation is reflected in no other way than the magnitude of the association expressing some kind of STR relative to other kinds.*

Some Illustrative Research Findings

Some illustrative findings will clarify the PCC, but before reporting them it is desirable to consider the context. Consider the numerous reports (e.g., Yip and Caine 2011) of an association between the unemployment rate (UR) and the suicide rate (SR). Despite several reports of a positive association, the findings are far from convincing evidence of causation, the immediate reason being variation in the degree of association (see Phillips 2013: 593 for a brief but recent review of the literature).

What is the most appropriate spatial entity for examining the UR/SR association (e.g., cities, states, regions, or countries)? Then what is the most appropriate time lag between the asserted cause (URs) and the effect (SRs)? When designing research, answers should be guided by the space and time implications embedded in the underlying assumptions about the causal process. If the process pertaining to rates is seen as *local and immediate*, we should examine the association between the URs and the SRs for cities or even smaller territorial entities, using data with the least lag time. Alternatively, if the causal process is seen as more complex and involving several common steps between unemployment and suicide in what are very mobile populations, the appropriate spatial entities might be states with a long time lag (Stack and Haas 1984).

Bear in mind that we are considering possible correlates of the SR, not the correlates of individual cases. Rabid reductionists in sociology to the contrary, you could have a theory about correlates of both. If so, *at least one* dimension of predictive power—range—would obviously be greater for a theory with multiple unit terms (i.e., two or more distinct parts, any two of which have different unit terms). "Unemployment" could be a variable in such a theory, as a *rate* in one of the two major parts and as an individual correlate in the other. Tests of the "individual part" of the theory would focus on the percentage of the unemployed who committed suicide and the percentage of suicide victims with an unemployment history. Note, however, that the predictive power of the "individual part" would likely depend on something akin to a contextual unit term (i.e., any pertinent percentages would necessarily refer to some larger population as the context, such as residents of a city vs. residents of a country), and there is nothing about recognizing parts of the theory that would preclude taking relative predictive power as the supreme criterion in assessing *all* parts relative to contenders.

In that connection note also that the unemployment/suicide relation as regards individuals is always some stipulated spatial context (e.g., "Among residents of any city"), and those who oppose so-called "ecological correlations" (actually "aggregate correlations") persistently ignore the possible importance of the larger spatial context of "individual correlations." Unfortunately, it is not just a matter of *particular* spatial contexts (e.g., the correlates of suicide cases in Japan vs. U.S. correlates). Additionally, there is the possibility of variation in the kind and magnitude of correlates, depending on the type of spatial context (e.g., urban vs. rural suicides).

The temporal relation between suicide and unemployment (or any other asserted correlate of suicide) is another difficult problem at both levels, individuals vs. populations. The only problem at *both* levels may appear to be discovering or deducing the strategic "time lag," that which maximizes the association between unemployment and suicide. That problem is paramount, but one of several complexities is the awful possibility that at both levels the strategic time lag is contingent on the spatial context. The authors know of no study confirming that awful possibility, but we would not be surprised by evidence that the strategic time lag (STL) between unemployment and suicide in the U.S. during the 1930s is markedly different from the current STL.

Needless to say, no social/behavioral scientist is likely to deny that suicide, whether individual cases or rate variation, depends on spatial and/or temporal context. So the real challenge in formulating a theory about suicide at any level is interpretation of evidence of space-time relevance. Thus, the study of suicide in the U.S. during the 1930s takes on special possible significance because of abundant evidence that annual rates of unemployment in the 1930s were much greater than for subsequent decades. What one makes of that particular "temporal contingency" is actually a step toward a theory about suicide, but there is a more immediate point. All data pertaining to suicide have a spatial/temporal context, and the theoretical significance of that context cannot be exaggerated.

Other Complexities

Even ignoring the distinction as to the two basic kinds of suicide theories (i.e., those dealing with rates vs. those dealing with individual cases), the choices are complicated enough. As one of many considerations, there are conditions akin to counterfactuals, such as the possible *absence* of prompt and effective medical care for self-inflicted wounds. Still another "contingent condition" is the presence of firearms in households (the American Rifle Association notwithstanding). Those conditions enhance the stochastic nature of the association between unemployment and suicide.

Such analytic complexities in the advanced sciences are commonly dealt with in carefully controlled experiments. So how about randomly assigning persons to experimental and control groups when investigating the possible links between unemployment and suicide? The very idea is so grossly offensive that we apologize for even introducing it, and we have done so only because it illustrates a distinctive limitation on social/behavioral science, granted some limitations for any science.

Kinds of Space-Time Relations in the Association of Rates

The foregoing led us to an analysis of several types of STRs between the UR and the SR, with States in the U.S. and years over 1976–1995 as the entities (all rates gathered online from annual official sources). Given what is reported below, we hope that interested readers will contemplate replicating our research as data are released for the states in the U.S. between 1996 and 2015.

A Typology of Space-Time Relations and its Significance

In analyzing the 1976–1995 data we employed the typology set forth in Table 4.1 (Chapter 4). Some of the types of STRs were considered along with subtypes distinguished by lagging variables, a total of seventeen types or subtypes.

Over recent decades such comparisons have become alien to social science research, with a turn instead to models, methods, and criteria of causal inference that implicitly deny or ignore the importance of distinctions as regards STRs. The denial commenced long ago, as witness Herbert Simon's statement (1952: 518): "We shall avoid the usual assumption that the asymmetry of cause and effect has something to do with sequence in time." Likewise, writing on causation more recently, two prominent social scientists, Pearl (2009) and Duncan (1966), seemed to have joined Simon in their apparent indifference to types of STRs as manifestations of causation, focusing rather on models and statistical method (e.g., path or regression coefficients).

Initial Findings

We found enormous variation in the product-moment coefficients of correlation (*r*s) between the URs and the SRs–annual in both cases. At one extreme, when the URs are lagged five years *after* the SRs the longitudinal *r* (i.e., Subtype IA2/5u, a subtype of Type IA2 in Table 4.1) is.840 for Ohio. At the other extreme, it was a negative *r* −.691 for Mississippi, which is for the *opposite* subtype (IA2/5s), meaning when the SRs are lagged five years after the URs.

In light of conventional thinking about causation and time sequence (Simon [1952: 518] and Blalock [1964: 10] not withstanding), the *r* of −.691 suggests that for Mississippi over the years in question, a relatively *high* UR for some year was a substantial cause of a relatively *low* SR five years later. No less contrary to conventional thinking, the *r* of.840 suggests that for Ohio a relatively *high* SR for some year was a substantial cause of a relatively *high* UR five years later.

Although the two coefficients,.840 and −.691, are extremely puzzling, they become less so on recognition that for any of the several types of STRs the UR/SR *rs* (hundreds of them) are, so to speak, all over the board, though *predominately* positive. But it would be a gargantuan mistake to think that such associational variation is peculiar to the findings in question. To the contrary, such variation is fairly common for sociological variables. And it would be another gargantuan mistake to think that some statistical method or technique precludes the aleatory or the fortuitous as regards possible manifestations of causation among sociological variables, whatever the STRs and whatever the kind of entities in questions—individuals v. cities, organizations vs, countries, or what have you.

Another Set of Findings

The means (averages) of the *rs* of the subtypes of STRs are much more informative than are the individual *rs*. In particular, the most extreme and *only* negative mean is −.045, that being for one-year absolute change in both UR and SR, with change in SR lagged one year after the UR change in this cross-state association for each of particular years (Subtype IIB2/1s). The other extreme is the positive mean *r* of.204, which pertains to the magnitudes (levels, not absolute or proportionate change) of UR and SR, with SR lagged five years after UR in that longitudinal correlation—each state over time (Subtype IA2/5s). Convincing or not, the mean *r* (.204) suggests that variation in UR may have been *a* cause of variation in SR for the entities in question (states and years), all the more in light of the mean *r* of.021 when the same lag is reversed (i.e., Subtype IA2/5u).[10]

Of course, one may balk at the very idea of a five-year "causal lag," but perhaps not on considering possible paths to suicide such as: UNEMPLOYMENT→LOSS OF CREDIT→MAJOR RESIDENTIAL MOVE→MARITAL DISRUPTION→LIFE ON THE STREETS →SUICIDE.[11] An even more relevant possibility is that an increase in the UR over several years is associated with longer periods of unemployment for the unemployed (see Stack and Haas 1984), perhaps so long in particular instances that the individual ceases to look for a job (such cases contribute to an evidential problem in that those who cease looking eventually are no longer counted as "unemployed").

Obviously, the difference between the first of the two crucial mean *r*'s, .204 and .021, is far from substantial, though the former is some ten times the latter, which is to say that the absolute vs. relative comparison is another complexity. Additionally, the standard deviation (SD) of the *r*'s that entered into the .204 mean is .332, the second greatest for all STRs (the greatest SD, .404, is for the *rs* that entered into the other crucial mean, that being .021). Of course, the mean of .204 might well have been much greater and the SDs much less had the SRs and/or the URs been, say, three-year annual averages. However, such rate distinctions are *not* something apart from *kinds* of STRs, and limited resources preclude a systematic consideration of kinds of rates (e.g., annual vs. three-year vs. five-year averages). For present purposes, the crucial point is simply that measures of association among sociological variables vary substantially from one type or subtype of space-time relation to the next.

Introduction of "Strategic Ratios"

In light of the extreme variation in the *rs*, the findings are more informative when analyzed in terms of "strategic ratios," defined as the mean of the *r*'s for some type or subtype of space-time relation *divided by* the SD of those *r*'s. The greatest strategic ratio, 0.6145, is for the longitudinal *r* (each of the 50 states over time) between UR and SR, with the SRs lagged five years (Subtype IA2/5s). Moreover, when that lag is reversed (the Subtype becoming IA2/5u), the strategic ratio is only 0.0520.

No less important than the lag contrast (i.e., Subtype IA2/5s vs. Subtype IA2/5u), consider the remarkable increases in the strategic ratios as the lag length *increases*: Subtype IA2 (no lag), −0.1240; Subtype IA2/1s, 0.2198; Subtype IA2/3s, 0.5858; and Subtype IA2/5s, 0.6145 (limited resources precluded consideration of other lag lengths, the expectation being that the *r*'s would commence decreasing). The pattern takes on additional significance in light of Phillips's report (2013) of negligible regression coefficients for the UR/SR association both across states in the U.S. and over time 1976–2000 for each state. The regression equations included a grab bag of other independent variables (e.g., percent foreign-born, population size, and percent Episcopalian), but that is secondary to the point that Phillips did *not* lag SRs or URs in the manner we did.

Does this pattern matter? Even though the strategic ratio for Subtype IA2/5s is over twelve times that for Subtype IA2/5u, is 0.6145 substantial? Any answer would be questionable, all the more in recognition that, assuming no *negative* mean of the *r*s, the absolute possible *minimum* for any strategic ratio is .000; but the *absolute possible maximum* cannot be computed, for the SD *could be* .000 (i.e., absolutely no variance in the *r*s). Nevertheless, the mean (.204) and the SD (.332) of the Subtype IA2/5s correlations alone preclude a more confident conclusion than something like this: for the U.S. over 1976–1995, variation in URs among states and over time for each state was at most only a negligible cause of variation in SRs. Lest that guarded conclusion suggest "Sonny, forget about it!", the findings *could have been* such as to indicate no support whatever for even a guarded conclusion.

The Bearing on the Second Dangerous Belief in Sociology

Stating the belief as an issue: even if defensible causal inferences should be based on data pertaining to STRs, neither the type of STR nor the kind of entities (e.g., countries vs. organizations vs. individuals) has any particular significance. It is difficult to imagine anyone denying that the previously reported findings on the UR/SR association sharply contradict that issue statement.

Of course, one can deny that the dangerous belief is widespread in sociology, but those who voice that denial are encouraged to meet this challenge: quote statements in the sociological literature that are in full *agreement* with this book's principal causal contention (*supra*) and clearly reflect *rejection* of what has been identified as sociology's second dangerous belief (*supra*). The literature search should include Duncan's paper on path analysis (1966) and Pearl's *Causality* (2009). Evidently, both publications have greatly influenced thinking about causation in the social sciences. Both authors wrote as though correlations do *not* necessarily pertain to some distinctive kind of space-time relation, let alone that causation may be manifested in any one of various kinds but more for some kinds than others depending on the variables. Should that claim be doubted, the naysayers are obliged to quote Duncan and Pearl at length.

Here we would simply note that Duncan wrote without mentioning Hume on causation, especially Hume's definition of a cause (1896 [1739]: 93): "An object precedent and contiguous to another, and where all the objects resembling the former are placed in like relations of precedency and contiguity to those objects that resemble the latter." Pearl does comment on Hume, even quoting (2009: 238) a slightly different version of Hume's causal definition; and Pearl's caustic rejection includes this remarkable statement (2009: 406) about Hume: "He knew quite well that the rooster crow *stands* in constant conjunction to the sunrise, yet it does not *cause* the sun to rise." Well, we must surely wonder how Pearl could have known what Hume knew, and in any case, upon closer examination Pearl's comment is risible. The sun has risen millions if not billions of times in uncountable places without a crow from any rooster, and so much for "constant conjunction."

The Relevance of "Specificity" as a Predictive Power Dimension

Is there no way out of the causal quagmire? Yes, and it has already been suggested in the treatment of predictive power, the "space-time specificity dimension" in particular (Chapter 4).

Recall that a theory has maximum specificity if the theorist stipulates for each theorem the kind of space-time relation that would supposedly maximize the *theorem*'s predictive accuracy.[12] Of course, each such stipulation is an assertion, and one that may prove totally false; subsequently, someone may demonstrate that some other kind of space-time relation results in much greater predictive accuracy.

Even so, the assertion is indicative of the theorist's belief about causal evidence for the variables in question. Should the theorist make her/his conception explicit, there would be no rational occasion for debate; and if one took place it would be sterile, though perhaps interesting. After all, the theorist's assertion would be limited to particular variables in a particular theory. To be sure, it could be challenged, but an objective challenge would have to be limited to what is claimed—that the variables will prove more closely associated for the kind of space-time relation stipulated than for any other kind. Mind you, specifying a kind must attend to several matters: (1) specification of the kind of entities, already given by the wording of the theory, if formally stated; (2) the exact nature of the temporal relation, including length of lags, if any; (3) the procedure for creating values of the variables; (4) the kind of data for application of the procedure; and (5) the kind of association measure (e.g., regression coefficients or some ordinal measure) to be employed in tests of the theory.

Taking the "specification strategy" seriously would do more than diminish sterile debates over causal evidence. It could substantially increase the frequency with which sociological generalizations or theories are commonly accepted as meeting the proposed causal criterion, which is to claim that the proposed criterion is not only empirically applicable but also more *realistic* than any other empirically applicable contender. The basis of the claim is that the proposed criterion makes causation a matter of degree or extent. When a theorist asserts that some particular kind of space-time relation exceeds any other (i.e., those pertaining to the same variables), he/she is not so much as suggesting "maximum association" or "no exceptions" (in the qualitative case).

But what happens to the idea of "the" cause? Sociologists should have abandoned it long ago! As a case in point, no one is likely to assume that variation in the UR is more than possibly *a* cause of variation in the SR. Indeed, it is entirely possible, indeed likely, that some other variable exceeds the UR when it comes to predicting SRs. Even so, the UR would not be abandoned in formulating theories about SRs, which is to say that the proposed treatment of causation is conservative in that it retains (even if tacitly) the realistic notion of multiple causation.

This reaction is anticipated: the "specification strategy" places an impossible burden on a theorist. How can he/she possibly know that some designated kind of space-time association between the variables in question exceeds all other possibilities? Well, it cannot be known with anything like certainty, but some theorists "guess better" than do others; and that talent should be honored. As for imposing burdens on theorists, sociology has suffered *ab initio* from "impositional leniency."

A theorist's burdens are less onerous than they appear. If you—the reader—try to formulate a theory about the UR/SR relation, you could do worse than assert a positive longitudinal association and stipulate a five-year lag between the two rates, with the SR lagged.[13] Of course, that stipulation would be ill-advised if based on only one study, but in formulating a theory searching the research literature is hardly a felony. Indeed, sociology has always suffered theorists who are poorly informed about the research literature.

Finally, the foregoing should be considered in light of this question: why is causation a thicket of problems and issues, especially for the social/behavioral sciences? Answering all too briefly, practitioners of other sciences have had the good sense to judge theories and generalizations in terms of predictive power (including and especially "specificity") and thereby largely avoiding the insoluble problems and unresolvable issues created by an insistence on demonstrations of causation. Even granting that social/behavioral phenomena are much more complex, why encourage social/behavioral scientists to take on problems and issues that can be circumvented?

The Principal Limitation of the Illustrative Research

Can there be any assurance that the distinctive strategic ratio for the Subtype IA2/5s relation between UR and SR would hold for, say, countries and cities? Not at all, and that answer relates to the principal limitation of the illustrative research—its restriction to states of the U.S. over 1976–1995.

But there is a larger and much more important issue at stake. Let us suppose that the strategic ratio in question (0.6145) approaches .000 when the UR/SR association is among cities, even U.S. *cities* over 1976–1995. If conventional thinking among sociologists is accepted, the finding for cities would promote this conclusion: Variation in URs cannot be even *a* cause of variation in SRs and the findings reported here for states over 1976–1995 are simply misleading, perhaps spurious. The point is that sociologists need to recognize that casual evidence is entirely a matter of context. Specifically, causal evidence in one context (e.g., a 3-year versus a five-year lag, or cities versus states) is not nullified by its absence in another context.

"Context" includes far more than types or subtypes of STRs or even the kind of entities considered in the research.[14] It also includes the way the values of the variables are expressed (e.g., annual vs. average annual rates), the nature of the data that entered into computation of the values, and the nature of the associational measure. Most controversial, it includes the *historical* time period (1976–1995 for the illustrative findings).

To appreciate the controversy, suppose that research reported here is replicated for U.S. states over 1996–2015 as data become available, and suppose further that the strategic ratio for Subtype IA2/5s approaches .000. To conclude that the difference precludes variation in URs as being even *a* cause of variation in SRs is tantamount to arguing that space-time manifestations of causation are *temporally invariant.* The parallel argument would be "also spatially invariant" (i.e., causation is manifested only when the data pertain to a particular kind of entity).

So can there ever be evidence of "*no* causal relation"? Of course, such evidence is entirely possible. Suppose, contrary to the findings just reported, there is *no* known context in which the strategic ratio for *any* type or subtype of space-time relation between URs and SRs is appreciably greater than .000. Those who believe there is a causal connection between URs and SRs would be hard pressed in defending that belief. For that matter, the belief would become pointless if there is evidence that some variable *other than* URs is in *all contexts* closely associated with SRs.

The proposed treatment of causal evidence is entirely consistent with the contention that generalizations and theories should be assessed entirely in terms of relative predictive power. If sociologists adopt predictive power as the supreme criterion for assessing generalizations and theories, nothing would preclude the possibility that in the distant future someone will identify a variable that, regardless of context, permits accurate predictions of SRs to the point that consideration of a contender would be a waste of time. Preposterous? Surely less preposterous than continuation of the epistemological anarchy that permeates sociological treatment of causal evidence, and the arguments made here are not limited to the UR/SR relation.

Final conclusion? Establishing causation is a vexed issue. Whatever else is involved, it depends very heavily on well specified STRs. Assuming otherwise will eliminate any chance of sociology becoming a genuine science.

Belief Three: Generalizations Are Unnecessary and Perhaps Odious

A third disastrous belief is that generalizations are unnecessary for science. To this the reply is straightforward. Unless based on sheer hunches, predictions require generalizations *beyond* particular events or things. Yes, measurement strategies to test generalizations are central to the scientific enterprise; but without generalizations, science scarcely exists.

For example, suppose some sociologist sets forth what he/she considers an explanation of the assassination of President John F. Kennedy. Suppose further that the explanation does *not* include any explicit generalizations about presidential assassinations, past or future. What we would have is an explanation (or interpretation) of a single event, not an assertion embedded in a theory or even a generalization about all similarly defined events. Yes, the explanation might suggest a generalization; but "suggestive/interpretive sociology" should have ended long ago, and adoption of formal theory construction would mark its internment.

This distinction between the explanation or interpretation of single events and assertions about an infinite category of such events is critical to doing science, and perhaps more than any other single factor the distinction questions the works of many sociologists (e.g., Wuthnow on Texas [2014]). Interpretations and explanations of particular events can be important, insightful, and persuasive, but they are not enough if we seek theories about how the world works.

A Classic Example

No sociologist better illustrates the point than Max Weber. His analysis of the relation between the Protestant religion and capitalism (Weber 1930), widely acknowledged as a masterpiece, must be one of the ten most cited works in sociology. However, it is also an explanation with no *explicit generalization* as to the conditions under which economic systems of countries or world regions change to "capitalism." If we are prone to defend Weber as a scientist, and there are numerous other reasons to do so, we might credit Weber for implying this generalization: Regardless of the location, when the Protestant religion thrives it will be followed or accompanied by capitalism.

Whether this generalization was part of Weber's thinking is debatable, but the adoption of capitalism by the Japanese is alone reason to question any interpretation of Weber in terms limited to Europe or even any particular century. In addition, subsequent scholars have noted that capitalism was thriving in Europe centuries before the Protestant work ethic was in place and that currently the wealthiest portions of Europe are as likely to be Catholic as Protestant (see Lopez 1976 and Stark 2005). Above all, Sanderson, *et al.* (2011) report findings from quantitative research on European economic variables over several countries that do not support Weber.

Despite reservations about Weber's theory, one could argue that Protestantism is a *sufficient* but not *necessary* condition for capitalism or some particular features of it. Specifically, wherever and whenever it is found, Protestantism involves a commitment to the future, hard work, and delayed gratification. Consequently, Protestantism is conducive to capitalism, or so the argument goes. However, contemplate Marx's (1963 [1855]: 109) materialistic generalization: "The hand-mill gives you society with the feudal lord; the steam-mill society with the industrial capitalist." The question then becomes: what criterion could yield *effective consensus* in sociology as to Weber vs. Marx on the origin of capitalism?

Whatever the eventual conclusions, the point here is that Max Weber appears to have been primarily interested in events that played out in the West and a rapidly changing Europe during two particular centuries. He did not focus on explicit *generalizations* about religion and economic systems across time and space.

Interpretive Sociology

Max Weber is often identified with what is commonly called "interpretive sociology." This perspective has a long *verstehen* tradition, including symbolic interaction, phenomenology, ethnomethodology, subjectivism, humanistic sociology, and dramaturgical sociology.

An Outstanding Case

In writing *Everything in its Path* (1976), Kai Erickson produced what became a famous model of interpretive sociology. His foremost goal was a description of the way residents of Buffalo Creek, West Virginia, responded to and recovered from a devastating flood. As he put it (1976: 246–247), "The flood on Buffalo Creek was a unique human event, a singular tragedy, and it resulted in wounds that can only be understood properly when they are seen in their own local context." However, he also alluded to "details that seem to be found in other events of a similar kind occurring elsewhere in the world."

The immediate point is that Erickson failed to develop an answer to this question: well, which of the "details" are found in "other events" and which are not? Had he answered that question, he would have approached one major goal of the scientific enterprise—the formulation of a generalization about an *infinite* category of events or things. Why did Erickson stop short of that goal? In his own words (1976: 12) "the aim of any science … must be to move from particular observations to general findings, but there are times when the need for generalization must yield to the urgency of passing events, times when the event must tell its own story."

Another Outstanding Case

The highly influential work of Erving Goffman and most particularly the idea of "stigma" (1963) led to numerous situation-specific accounts over the years, including interpretations of spoiled identities in cases of autism (Gray 1993), becoming a felon (Uggen, *et al.* 2004), the presence of physical or mental differences (Kayama and Haight 2014), being gay (McRuer 2006), and many more. Along with "stigma," in *Asylums* (1961) Goffman developed what can be rightly described as "explaining by naming," wherein we are given descriptive accounts of the characteristics of total institutions, moral careers of mental patients, and the underlife of public institutions.

Although Goffman was highly influential, he is hardly the only adherent to the "explaining by naming" tradition. Hannah Arendt (1963), who, while not strictly a sociologist, is frequently so referenced, provided the "the banality of evil" label to capture how Adolf Eichmann transformed from a "terribly and terrifyingly normal" German citizen into a Nazi war criminal. Subsequently, the banality-of-evil label received considerable attention, and it was used recently to interpret the notorious events at Abu Ghraib prison during the second Iraqi conflict (Bernstein 2008).

Still Another case

Another outstanding instance of interpretive sociology is Eric Klinenberg's (2002) compelling "social autopsy" of the July 1995 Chicago heat wave, one that claimed the lives of some 700 persons. Klinenberg explains his approach as follows (2002: 18):

> Just as the medical autopsy opens the body to determine the proximate physiological causes of mortality, this inquiry aims to examine the social organs of the city and identify the conditions that contributed to the deaths of so many Chicago residents that July.

Klinenberg was well aware of existing disaster literature and used it to enrich his social autopsy of the Chicago heat wave, drawing interpretive insight from a wide range of studies, such as Erickson's (*supra*). Like Erickson, however, Klinenberg's major purpose was to shed light on a single human event.

For Klinenberg, there were patterns in the consequences of the heat wave, patterns created by differences among areas of Chicago as regards social isolation, social inequality, and pervasive poverty. These patterns were made all the worse by "an aging population of urban residents who live alone, often without proximate or reliable sources of routine contact and social support" (Klinenberg 2002: 230). With this situational interpretation in hand, Klinenberg's attention turned *not* to formulating or testing generalizations about disasters as an infinite category of events. Rather, he posed policy issues, among which were remedies for the mismatch created by (1) the increasing "delegation of key health and support services to paramilitary governmental organizations," (2) "the expectation that city residents, including the elderly and frail, will be active consumers of public goods," (3) the increasing "social distance between city administrators and the disadvantaged people they serve," and (4) the increasing use by governments of "public relations and marketing programs to spin good news stories about the success of their programs" (Klinenberg 2002: 231–232).

From Klinenberg's interpretive study of Chicago's 1995 heat wave tragedy at least two testable generalizations are suggested. First, within neighborhoods, the greater the social isolation of identified groups (e.g., those over 65 years of age), the greater the harm experienced by these groups when disaster strikes. Second, the greater the poverty of identified neighborhoods, the higher the death rates experienced in those neighborhoods when disaster strikes.

Bottom Line

In each instance—Weber, Erickson, Goffman, Arendt, and Klinenberg—an account provides an interpretive framework to better *understand* particular situations in particular places and at particular points in time. While laudable for any number of reasons, the reluctance to move beyond situation-specific interpretations to testable generalizations will not lead sociology toward a robust body of theories.

The path to a stronger scientific standing of sociology lies in rejecting the belief that "generalizations are not essential." Alternatively stated: Whatever an explanation may be, it is the *expressed or implied* generalizations that are the most relevant for science. There is only one qualification: consistent with everything said in previous chapters, the merits of any generalization are judged as a matter of its predictive power.

Quantitative Situation-Specific Findings

Do not construe the foregoing as suggesting that Max Weber and other interpretive sociologists are alone in writing with scant attention to generalizations. They are not. Substantial time and resources are spent every year by survey researchers and demographers simply describing particular events, settings, or trends with no concern whatsoever for theoretical underpinnings. With this tendency noted and the present carping notwithstanding, it should be said that careful attention to specific details in particular places and times is also essential to the scientific enterprise.

Single Data Points

On October 7, 2014, the Pew Research Center (PRC) reported that the share of affluent students who borrow money for college had doubled over the past decade. In 2015, drawing from their 2014 Religious Landscape Study, the PRC also reported that the percentage of persons identifying as Christians fell from 78.4 percent to 70.6 percent between 2007 and 2014.[15] Likewise, demographers routinely report a wide range of factoids on the estimated number of immigrants from and to places around the world. In particular, the World Bank compiles a Global Migration Database, indicating that global migrant stock increased from 92 to 165 million between 1960 and 2000.[16] And so it goes.

Once again, we should be quick to clarify. The label "factoid" should not be construed as dismissive. Call them what you will, they could play a crucial role in science. It is from such data that uniformities can be discovered and generalizations formulated, the road to genuine scientific theories.

Stepwise Regression

The sociological literature is filled with multivariate equations that have few, or very poorly specified, theoretical foundations. Some of the related issues are addressed in Chapter 7. Here we limit attention to what is known as stepwise regression.

The procedure commonly commences with an outcome variable that the researcher wants to better understand or predict, such as success in the first year of college. In the typical data set there are numerous variables, such as SAT and/or ACT scores, rank in high school class, parents' income, a score for writing competency, where the high school was located (rural, urban, or suburban), a quality indicator for high school curriculum, how many and which math and science courses a student has taken, whether parents had attained a college degree, some measure of extracurricular activities, etc.

Not having an underlying theory to test or build upon, the researcher proceeds by adding variables one at a time to a regression equation, beginning with one that yields the greatest "explained variation." Step by step he/she will then proceed by adding variables until they reach a point where there is no appreciable (as determined by some algorithm) increase in "explanatory power." Alternatively, the researcher may proceed in the reverse direction, beginning with all available variables entered into the equation. He/she then deletes variables one by one, comparing the stepwise results to some standard, until concluding that further deletions do not improve the efficiency of the model being investigated.

Such procedures might yield interesting results for a particular situation, but that is all. Interpretation of the progressively attained regression coefficients and explained variation is necessarily situation-specific, similar in this regard to community ethnographies and historical accounts of specific events. Were we to repeat the same procedures with measures of the same variables in another setting or in the same setting at a different time, we might get quite different results.

There is, of course, much to say in support of carefully collected data points, whether qualitative or quantitative. Given the now mountainous amount of data available, social scientists need to gather these data and focus their efforts—as did Copernicus, Galileo, Brahe, Kepler, and eventually Newton—to formulate overarching, testable generalizations. Equations may be essential steps toward generalizations, and eventually it may be desirable—perhaps necessary—to state the generalizations in the form of equations (with attached unit terms). Nevertheless, unless a generalization is the goal, the activity is dustbowl empiricism, no matter the technical sophistication of the researcher and the equations produced.

Situation-specific interpretations or assessments, whether qualitative or quantitative, are necessary but not sufficient for any science. When they are used to expand the predictive power of generalizations and theories, we begin to move with the aid of a much-needed map. Nonetheless, without generalizations and theories to guide our pursuit of knowledge, we are much like the proverbial prophet wandering in the wilderness.

A Final Take

This brings us to a final point. Generalizations about a finite category of events or things are of less scientific significance than generalizations regarding an infinite category (i.e., unlimited spatially or temporally) of such events or things.[17] Consider the *greater* average annual U.S. crude birth rate (CBR) for 1920–1929 than for 1930–1939. Suppose the explanation reduces to this assertion: the CBR declined because of an increase in the UR over the two periods. When the logic of induction is applied to the assertion, we arrive at this infinite but testable generalization: for any country and any two similar periods of years, the average annual CBR is greatest for the period having the lowest UR.

Whatever the outcome of subsequent tests, one dimension of predictive power—range—is much greater for the infinite generalization. Although the predictive accuracy of the infinite generalization cannot be even estimated without several tests, the very notion of predictive accuracy applies only in a very restricted sense to any assertion, whatever the variables or attributes, that is limited to one time point or period for some particular entity, the U.S. or otherwise.

Final point: it is not enough for sociologists to agree that generalizations and theories are essential for science. They must also recognize that those "essentials" will not be realized until the disastrous beliefs examined in this chapter are abandoned. We end this chapter with this: you can make book on the last contention.

Notes

1 http://www.norc.org/NewsEventsPublications/Pages/published-research.aspx/
2 http://www.bls.gov/nls/

3 http://www.dhsprogram.com/
4 http://www.census.gov/acs/www/
5 https://cps.ipums.org/cps/index.shtml/
6 http://www.cessda.net/
7 http://www.nimh.nih.gov/health/publications/brain-development-during-childhood-and-adolescence/index.shtml/
8 http://hraf.yale.edu/resources/guides/basic-guide-to-cross-cultural-research/#selection-of-cultures/
9 http://tylervigen.com/spurious-correlations/
10 Such judgment reflects a criterion of causal evidence, and space limitations permit only a brief statement of it, one limited to bivariate quantitative variables. Variation in *X* is a cause of variation in *Y* to the *extent* that (1) changes in *X* values are statistically associated with a later change in *Y* values and (2) changes in *Y* values are *not* statistically associated with later changes in *X* values. The criterion is consistent with Hume's treatment of causation, though he did not sufficiently recognize the relevance of types and subtypes of space-time relations, including "lag reversals." Moreover, the criterion is contrary to this principle: a causal connection, if any, is *not* manifested in the same kind of space-time relation for all variables and all kinds of entities.
11 Such possibilities make the treatment by Hedström and Ylikoski (2010) of "causal mechanisms" relevant. Unfortunately, however, they (1) treat the subject as though problems and issues concerning causal evidence—especially the relevance of different kinds of space-time relations—can be ignored and (2) all too readily (p. 51) dismiss the suggestion "that a mechanism is just an intervening variable…"
12 In the case of isolated generalizations, unless the generalizer stipulates otherwise, it is assumed that the type of space-time relation maximizes the association in question.
13 The most strategic kind of entity for such a theory remains an open question. In any case, be warned that there is already a contending theory in the literature, one with considerable scope in that it asserts several correlates of the SR *in addition to* the UR. That contender is the status integration theory (see Gibbs 2001 for pertinent references). It implies a positive association between the UR and the SR because the status "unemployed" has a low degree of integration with any other status (e.g., "female-age 34"). Status integration is simply the proportion of people occupying one status, ascribed or achieved, who also occupy a particular achieved status. Thus, the theory implies a surprising prediction for *some* countries: The suicide rate of "married, ages 15–19" is greater than that of the "single, ages 15–19." Why? Because in some countries the percentage of residents 15–19 who are single greatly exceeds the percentage married. The theory also implies that in all countries beyond some age the suicide rate declines. Why the prediction? Because knowing that someone is, say, age 95 enables you to confidently predict that he/she is widowed and not in the labor force, which is to say a high degree of status integration. Of course, such implications are not meaningful without recognizing that the theory links status integration with the stability and durability of social relations.
14 "Entities" (designated by unit terms) are of special significance if only because sociologists have devoted little attention to this question: For any two variables, why does their association depend on the entities being compared? Anyone who doubts that contingency knows little about the findings of sociological research.
15 See, for example, http://www.pewsocialtrends.org/
16 http://databank.worldbank.org/data/views/variableselection/selectvariables.aspx?source=global-bilateral-migration/
17 Needless to say, whether any category—finite or infinite—is "sociological" depends entirely on the way the field's subject matter is delimited. Understandably eager to avoid the appearance of pretending omniscience, the authors admit that this book scarcely has any bearing on the delimitation of sociology's subject matter. However, we do grant that delimitation is one of the field's major problems. Sociologists study such a variety of events and things that the field's subject matter defies an empirically applicable or even comprehensible definition. The only hope lies not in delimitation of subject matter by fiat, no matter how empirically applicable; rather, only defensible theories will point the way, which makes criteria for assessing theories all the more relevant.

9

THE QUEST FOR UNIFORMITIES AND PROPOSITIONS

With compliments to readers who have persevered this far, we turn to a final subject. As suggested at various previous points, whatever the field or branch, science is a search for uniformities. When detected, uniformities are commonly expressed as propositions, each of which potentially implies predictions, or should do so. Any such prediction necessarily pertains to particular sets of events or things, and science requires preponderantly accurate predictions to further entertain the uniformity asserted by the proposition.

Of course, considered as a body of knowledge, science is more than a conglomeration of propositions. Ideally, propositions would not stand apart but, rather, united by deductions from a particular set of premises. But propositions may come from any number of sources including sheer intuition, a feat of imagination in which possible uniformities are detected without direct observation. To repeat, there is no logic of discovery in science. As for the survival of a proposition or a theory, it *should be* determined by its relative predictive power, which is to say that there is or *should be* a logic of justification in science. The normative emphasis simply recognizes that contemporary sociology is a far cry from genuine science.

When speaking of genuine science as manifested in uniformities and propositions, a name rapidly disappearing from sociology's collective memory is George C. Homans (1910–1989). In *The Human Group* Homans (1950: 115) asked: "What single general proposition about human behavior have we *established*?" The answer, he concluded, was elusive. Homans aimed to locate possibilities through an "intensive and systematic scrutiny of five small groups," believing that his analysis would result in general propositions reaching beyond the particular groups studied to data-based generalizations with broader implications.

A decade later, in *Social Behavior: Its Elementary Forms*, Homans continued his quest for systematically organized generalizations, declaiming (1961: 11):

> We propose to show how a number of the better established empirical propositions about social behavior may be derived from, and thus explained by, a relatively small number of general propositions applied to a variety of commonly occurring given conditions.

Homans was aiming for "a more adequate intellectual organization of the field," but he recognized that he stood at virtually the starting point. His conclusions included systematic deductions, symbolic representations, and mathematical formulas; but he was quick to say "The general propositions we shall use can only at present be stated with far less mathematical precision than [Isaac] Newton's." Instead, "explanation sketches" rather than "logically watertight explanations," should be pursued. Accordingly, Homans presented arguments (1961: 11) "suggesting how the empirical findings might follow from the general propositions under conditions we can take as given."

Homans' pioneering work drew considerable attention, but many sociologists viewed him as guilty of the ultimate crime in the field—reductionism. In any case, he did not survive the social construction of reality (e.g., Berger and Luckman 1966), the situation-specific interpretations of phenomenologists (e.g., Schütz 1967), ethnomethodology (Garfinkel 1967), and various versions of postmodernism, all pushing what came to be known as epistemic relativism (e.g., Luper 2004). In the years following Homans, a few sociologists pursued syntheses of data-based generalizations (e.g., Blalock 1967; Black 1976, 2011; and Sewell, *et al.* 2001); but they did not engage in formal theory construction as conceptualized in this book. Nor did Homans, but he stands center stage in sociology's history when it comes to emphasizing the importance of uniformities and propositions.

More on Uncertainty in Science

Although scientists attempt to reduce uncertainty through the predictive power of their assertions, some uncertainty will always remain, which is why we write about assertions rather than facts (contra Wootton 2015, esp. 259–301). In particular, epistemic links between concepts and data are in the final analysis debatable and measurement errors are inevitable. While the problems may be minimal in some sciences, they loom large in others, especially sociology. However, whatever the field, when the uncertainty of measurement is embedded in the investigation of stochastic processes, the predictive accuracy in testing propositions or theories is frequently far from impressive.

Uncertainty should be taken as a given in sociology, but it is also quite evident in psychology. With a view to assessing the extent of uncertainty in psychological findings the "Reproducibility Project" was organized as "an open, large-scale, collaborative effort to systematically examine the rate and predictors of reproducibility in psychological science" (Nosek 2012: 657). Five indicators of reproducible findings were developed and the principal conclusions were:

> Ninety-seven percent of original studies had significant results (P <.05). Thirty-six percent of replications had significant results; 47% of original effect sizes were in the 95% confidence interval of the replication effect size; 39% of effects were subjectively rated to have replicated the original result.
>
> *(Nosek 2012: 657)*

Given the inherent uncertainties and modest reproducibility of asserted uniformities, for many the questions become: "Why bother? Why not simply describe what we see as well as we can and be done with it?" Admonitions follow: forget about reproducibility and the search for predictions based on discovered uniformities; they are not worth the paper they are written on. We gain more insight from the accounts of policy wonks, clergy, historians, novelists, journalists, and poets than we can ever hope to secure from social science research or theories.

Given the current state of sociology, it is difficult to argue successfully against such admonitions. Nevertheless, that is what we are doing in this book. Why? The accomplishments of science in numerous fields are simply too impressive to ignore, though what we have been advocating is far more a hoped for future than a celebration of accomplishment.

An Illustrative Uniformity and Complexities

Rather than continue in abstract terms, an example may help. Over several decades sociologists and criminologists have studied variation in the homicide rate (HR). Their findings have consistently suggested a positive association between a population's HR and the amount of income inequality within the population (see Wilkinson and Pickett 2009 and Tcherni 2011 for recent literature reviews). That association holds for large cities, metropolitan areas, states, and countries. So, it surely qualifies as a uniformity, though the degree of association and the time range are another question.

Despite impressive findings, the association is not beyond question. For one thing, there is no evidence of a positive annual association for the U.S. as a whole during particular periods (e.g., 1970–2010). Then a major issue is posed by two other well-known positive correlates of the HR among some U.S. territorial entities (e.g., metropolitan areas)—income measures construed as indicative of poverty (including average or per capita income) and percent of residents classified as "black" or African-American (see Tcherni 2011 for a literature review and original research on U.S. counties). The issue is best stated as a question: Which of the three—poverty level, income inequality, or percent black—is the "real cause" of variation in the HR? In attempting an answer, investigators have been much more concerned with statistical techniques than the formulation of a theory. But consider the possibility of two theories about variation in the HR, one in which the major independent variable relates to income (poverty or inequality) and the other to percent black. Who would doubt that in extensive tests, especially at the international level, both the predictive accuracy and the range of the "income theory" would vastly exceed the "racial theory"?

Those differences in dimensions of predictive power might not hold in the case of an "income inequality theory" vs. a "poverty level theory," but currently the pertinent findings are far too limited to make an invidious comparison. For that matter, current practices notwithstanding, the real challenge would be a synthesis of the two theories, one in which the values of the two variables are *not* somehow combined by some statistical technique without a theoretical rationale. Rather, the synthesizing theory would be such that the two positive associations with the HR—one with income inequality and one with relative poverty—would be deduced from a set of axioms and postulates or postulates alone. Needless to say, the same strategy would apply in considering what appears to be contending uniformities or propositions, whatever they may be, perhaps two quite different correlates of some particular kind of rate of human fertility.

Some Crucial Qualifications and Complexities

In light of previous considerations, we would say that the range of the income/homicide association is limited, meaning the evidence suggests that the association is not equally robust for all kinds of entities (including years). The job of those investigating this particular asserted uniformity thus becomes that of identifying strategic contingencies. It may be a matter of context—U.S., European, or Japanese. It may be a matter of entity size, as anyone who has analyzed HR's extensively knows that the *annual* HR for very small spatial entities (e.g., all Ohio cities of, say, less than 5,000 residents as of, say, 2010) commonly fluctuates enormously, and that fluctuation is virtually certain to further the problematical character of the association between some income variable (IV) and the HR.

Detailed comparative studies may help us better understand the significance of the European, Japanese, and U.S. differences, though in the case of small spatial entities it may be a simple matter of using average annual rates for multiple consecutive years. The point is that investigations of exceptions to the IV/HR association are as important as instances of predictive accuracy. Such a strategy helps us refine assertions and thereby improve predictive accuracy, though perhaps reducing the range of the generalization in question.

We have often taken special note of the importance of unit terms. The point is not just that any asserted association between or among variables implies a unit term (e.g., age groups, families, cities, occupations, or countries). Additionally, in asserting a uniformity we must carefully specify the *kind of entities* to which the assertion applies. Otherwise, the asserted uniformity is vulnerable to a premature dismissal.

Situational Assessments versus Uniformities

It comes to this: asserted uniformities and tests of them form the foundation of science. Although limited in both space and time, situational assessments can lead to the detection of possible uniformities in an infinite category. Reconsider Erickson's (1976) account of a single community's response to a devastating flood and Klinenberg's (2002) "social autopsy" of a heat wave disaster in Chicago. These situational assessments did not give rise to a theory, but they did create the potential for one.

Situational Assessments and Value Judgments

Situational assessments, whether statistical or ethnographic, are currently far more common in sociology than are testable, theory-based assertions of uniformities *across* situations. One possible reason is that much of sociological research is driven by value judgments rather than theory-related empirical concerns.

A Possible Enormous Expansion of Science's Boundary

Withhold immediate judgment as much as possible while seriously contemplating four statements. First, one supreme moral action is to reduce human pain and suffering. Second, do unto others as you would have them do unto you. Third, violations of laws are never justified. Fourth, religion and government should be clearly separated in all human activity.

Whether such statements are purely normative and whether there is agreement or disagreement concerning their substance, there is a widely accepted contention that they are beyond the purview of science. The foremost rationale for their exclusion is the widespread belief that science deals only with "facts." Granted that this perspective is dominant among scientists, it could be challenged if "assertions" are taken as science's subject matter, as is the case here. Specifically, while it is difficult to understand how any normative or value declaration can be regarded as making any "factual" claim, it is no less difficult to imagine a defensible denial that such declarations are assertions.

The argument turns on the connection between "predictive implications" and "testability"; and it is our hope that Chapter 4 has made it clear that tests pertain to actual events or things, which is to say they are "empirical." Unless scientists develop some version of deontic logic for *particular* purposes, the predictive implications of normative/value assertions will remain obscure. Accordingly, if scientists develop that logic and take assertions as their subject matter, it could expand science's boundary enormously. The enormity of the challenge should be obvious, but even attempting to meet it is something far beyond this book.

Our position is that the research of sociologists has no necessary connection whatsoever with *some* value judgments and hoped for outcomes, though the findings may play a critical role in future developments. As a case in point, what is commonly referred to as deontological ethics, valued outcomes are anchored in moral obligations or duties regardless of consequences. An example is a belief system grounded in the duty to protect life and the accompanying correlates that life begins at conception, ending only when all brain electrical activity ceases. Policy conclusion: abortion should be banned, as should physician-assisted dying, no matter the consequences. In this deontological "pro-life" framework, social science is of no use whatsoever. Anticipating and assessing consequences are off the table, and deciding the goodness or badness of some action is a matter of values, not an empirical judgment.

Taking the second route to value judgments, *consequences* are given primacy. Partisans develop policies with an eye on anticipated outcomes, judged desirable or not. Social science is of no use when determining the desirability of outcomes, but it is central when attempting to anticipate actual outcomes.

Assume we value justice. Various indicators clearly suggest that practices embodied in Jim Crow laws (Woodward 1955; Foner 1988) left a legacy of unequal allocation of life opportunities across generations. "Affirmative action" policies are implemented to address the perceived injustice. Yet some of these policies impose what critics see as an injustice of a different sort and perhaps question even the anticipated outcome of the policy. Again, social science stands mute on what constitutes a just or detrimental outcome, but it is critical when it comes to actual consequences (e.g., Bowen and Bok 1998; Orfield and Kurlander 2001).

As the foregoing suggests, policy makers are frequently confronted with competing value judgments. To cope with the resulting dilemmas, priorities are set and choices made; and we commonly find arguments that the end (equal access to life's opportunities) justifies the means (temporary injustice while the system adjusts). Closely related, we encounter the doctrine of double effect, the idea that we take actions to achieve an outcome thought to be a greater good, knowing harm will follow.[1] This doctrine is at the heart of the debate over "death with dignity" laws. There is a moral obligation to alleviate suffering and a competing duty to protect life. Death with dignity laws underscore the good (alleviation of suffering) knowing harm (death) will follow.

Social scientists can gather data on how the process works and whether the adopted policy or law results in the anticipated consequences (e.g., Hedberg, *et al.* 2009). It is a grievous mistake, however, to think that beyond identifying predictive implication and possible actual consequences, social science can dictate priorities or value judgments.

Politics and Situational Assessments

Assuming social science's focus on the world as it "is" and not on how it "ought to be," we underscore an important point: sociology is simply not up to the task of providing theory-based predictions, the very thing policy makers need to craft policies with real confidence that the valued outcomes will be realized. For that matter, in light of current politics, uncritical acceptance of sociological findings could lead to catastrophic policies and laws. As for theories, it may well be that sociologists can never formulate the kind used by engineers building bridges. Indeed, in light of Stalin's Russia, Hitler's Germany, Mao's China, and the lesser-known consequences of the eugenics movement (late nineteenth and early twentieth centuries), there is widespread opposition to anything that smacks of "social engineering."

Instead, policy makers create policies and enact laws loosely guided by value perspectives, sometimes deontological sometimes consequential, ranging across the political spectrum. More specifically, they seek answers to questions such as: What is the gender gap in pay for the same work? Are there disparities across ethnic groups as regards to health care? What is it like to live on the margins of community life? To what extent are there fraudulent filings in Medicare claims? How do recipients use food stamps? Do current admission policies at universities comply with Supreme Court decisions? Are the "Death With Dignity" laws in Oregon, Washington, Vermont, and California yielding the desired results? Does capital punishment reduce the murder rate? To what extent are we executing innocent persons? Does the execution of an offender have a healing effect, perhaps bringing "closure," for the victim's family? What is the relation between labor unions and income inequality?

The list goes on and on, but it need not be extended in order to make a crucial point. Laws and policies invariably reflect assumptions about uniformities. Sociologists can justifiably question those assumptions, and perhaps even assert contrary uniformities. However, to so function sociologists are obliged to be empirically well informed, and as scientists they are also obliged to state asserted uniformities as propositions, if not clearly formulated theories.

Enacting laws and crafting policies remain very much a patchwork of value judgments, interest group influence on anticipated consequences, and the personal biases and commitments of policy makers. Whatever the value perspective taken or the issue at hand, there are calls for assessments of current circumstances to shape a policy or law; and if policies or laws are already in place, assessments of their efficacy. So it is that value-laden beliefs limit if not determine the empirical questions sociologists are likely to ask.

So what does policy-driven research imply for the prospects of developing formal theories? "Dim" would be the most appropriate response, but there is hope. Research driven by this or that policy issue is by its very nature focused on specific situational assessments; consequently, there is substantially less emphasis on trans-situational uniformities. Once gathered, however, situational assessments may encourage the search for uniformities, and therein lies the hope.

With this ray of hope, why do we assess the prospects as "dim"? Interest groups guiding the pursuit of specific issues, such as the gender gap in wages in the U.S. in 2015, are highly influential in determining funding patterns and therefore the nature of the questions/problems that sociologists pursue. Funding patterns alone may account for why, with very few exceptions, over the past several decades there has been an inordinate amount of situational assessment and very little attention given to the quest for uniformities, let alone general theoretical issues/problems. As long as situational assessment remains the primary driving force for funding sociological research, the quest for uniformities will not be emphasized and sociology as a science will suffer.

Particular Situational Assessments as Examples

It is important to note that the influence of policy-driven social science is not always anticipated or recognized. In the not-so-distant past, social science was used (though eventually rejected) to justify mandatory state sterilization laws, legislation eventually approved in 1927 by the U.S. Supreme Court. Defenders of the legislation argued that mandatory sterilization of those considered "feeble minded" or otherwise "unfit" would improve the nation's public health (see, e.g., Lombardo 2008). These efforts, organized by the Eugenics Record Office in Cold Spring Harbor, New York, gave rise to the empirical foundation for innovative policies and laws designed to minimize the perceived "threats" to public health. Far from confined to the U.S., such policies eventually provided a foundation for similar policies and laws in Nazi Germany (commencing in the early 1930s), thereby planting the seeds for what came to be known as the Final Solution (E. Black 2003; Ekland-Olson 2012: 13–92).

On another front, Gunnar Myrdal's *An American Dilemma: The Negro Problem and Modern Democracy* (1944), left a lasting legacy and promoted a decline in racist laws and public policies (see, e.g., Southern 1987, Clayton 1996).[2] On the other hand, Daniel Patrick Moynihan's *The Negro Family: The Case For National Action*, released some twenty years later (1965), was greeted with mixed reactions (including considerable hostility in sociology) that eventually led to a restrictive atmosphere as to what should and should not be investigated.[3]

In the years immediately following Moynihan's publication an atmosphere of "forbidden" questions developed among some social scientists (see *Annals of the American Academy of Political and Social Science* [2009]). Such developments influenced social science research in the following decades, becoming dramatic in the intellectual furor that came to be known as "The Bell Curve Wars" (e.g., Fraser 1995) and charges that "political correctness" had taken over the social science agenda. Less recognized, the issue really centered on assertions of uniformities and their interpretations.

More recently, political influence on the questions asked and answered by social scientists is evident in the acrimonious debate over same-sex relationships and marriages, with special attention given to the well-being of children raised in these relationships (e.g., Pawelski, *et al.* 2006; Regnerus 2012).[4] For some factions, if the political implications of empirical findings on some question go against this or that value perspective, then that question should not be raised, let alone answered. Such findings and questions readily become "forbidden knowledge" (Shattuck 1996).

For a discipline engaged in seeking information about the world as it is, the foregoing is truly alarming. As to a secure scientific standing, sociology must accept the "is/ought" distinction and produce an unbiased assessment of the "is" as much as possible, regardless of the prevailing political climate and pressures. As Sara McLanahan and Gary Sandefur argued (1994: 2) when studying what it meant to grow up with a single parent:

> We reject the argument that people should not talk about the negative consequences of single motherhood for fear of stigmatizing single mothers and their children. While we appreciate the compassion that lies behind this position, we disagree with the bottom line. Indeed, we believe that not talking about these problems does more harm than good.

Were it not for the frequently encountered constraining influences on social science to the contrary, the point of the quote would hardly be worth mentioning. However, politics being what they are, it is important to note that social science questions have been highly influenced by the politics of the moment. This influence has not always been in the spirit of open inquiry, and therein lies what is perhaps the greatest danger.

Some Pressing Needs

Whatever the sources driving the questions being asked, if sociology is going to succeed as a science, careful attention must be given to conceptual problems and issues. For example, in the foregoing chapters we have alluded to situational assessments of capitalism. How might they be used to formulate testable theories?

We first need agreement as to what we are looking for. If seen as a qualitative phenomenon, what are the properties and attributes of capitalism? If quantitative, what are its dimensions and how do we measure them? To what extent are particular religious beliefs especially correlated with capitalism? What are distinctive mortality and fertility rates, if any, of capitalist countries? What is the historical and contemporary association between labor unions and capitalism? Extending Thomas Piketty's already widely regarded work (2014), to what extent is extreme income inequality a *unique* correlate of capitalism? Does capitalism give rise to or promote racism and gender bias?

Note especially that all such questions pertain to the quest for uniformities. However, whether capitalism or something else, there is a need for conceptual clarification and agreement on measurement strategies. Conceptual clarity and consensus are necessary conditions if sociologists are to compare situational assessments with a view to detecting uniformities and devising tests of the related propositions. In light of the seven dimensions of predictive power described in Chapter 4, we have a greater sense as to why formal theory construction is not for the faint of heart. Again, an example may help.

A Particular Theory

Two dimensions of the predictive power of a theory are its scope (number of dependent variables) and its range (applicability to various types of units across space and time). Along these lines, the decades-long work of Donald Black (1976, 2011) is instructive. As noted previously, there can be no doubt about the ambitious range and scope of Black's assertions. Defining "law" as governmental social control, in *The Behavior of Law* Black writes (1976: 3–4):

> The quantity of law varies in time and space. It varies across the centuries, decades and years, months and days, even the hours of a day. It varies across societies, regions, communities, neighborhoods, families, and relationships of every kind … It varies across the world and its history … All of this is the behavior of law, and it is possible to explain all of it.

Similarly, as noted in Chapter 5, in *Moral Time* Black (2011: xii) writes that his theory "pertains to conflict in literally every kind of human relationship, whether between relatives, friends, colleagues, strangers, groups, or nations." Continuing, Black argues that his theory orders the findings described by practitioners in sociology, criminology, anthropology, and history, thereby explaining "why conflict differs from one social location to another, and why it changes over the course of human history" (2011: xiii).

This is not the place to assess Black's ideas in detail, which others have done in several venues (e.g., Sherman 1978; Gottfredson and Hindelang 1979a, 1979b; Greenberg 1983; Hunt 1983; Hembroff 1987; Lessan and Sheley 1992; Wong 1998; Bernard 2002). Instead, we briefly address questions related to conceptual clarity, the logical structure of ideas, and the relevance of the dimension's predictive power explicated in Chapter 4.

Black's Generalizations

Reviewing *The Behavior of Law*, Greenberg (1983) praised Black's theoretical framework, noting how Black had reduced findings from numerous empirical studies to a small set of bivariate assertions, such as: (1) "Law varies directly with rank … People with less wealth have less law. They are less likely to call upon the law in dealing with one another" (1976: 17) and (2) "Law varies inversely with other social control" (1976: 107). The same can be said of Black's *Moral Time*. In this sense Black's work fares well in light of the arguments made in previous chapters, especially the explicit character of his generalizations and their range/scope.

Where Black's theorizing has been most severely challenged is in regards to the clarity of his concepts and the specification (or lack of) of links among the components of his theory. As Greenberg argues (1983: 339) "Black's 'theory' consists of logically unrelated propositions asserting a relationship between pairs of variables" (1983: 339). But the identified dimensions of social life—stratification, morphology, culture, social organization, and social control—may be operating simultaneously. If so, making the *interconnections* explicit would increase the accuracy of whatever predictions are deduced from Black's generalizations.

Perhaps most obviously, Black frequently makes bivariate assertions that assume *ceteris paribus*. Accordingly, if the dimensions of the "equalizing conditions" are not specified, then the negative test outcomes can be explained away all too readily. Even ignoring that point, there is another, one best expressed as a question: are the variables clearly defined and are their linkages measurable? For reasons subsequently suggested, a clear-cut and defensible answer is difficult.

Testability and Conceptual Clarity

Again, the propositions in *The Behavior of Law* are largely bivariate and strictly associational. However, causal implications do creep in here and there. For example, Black (1976: x) suggests that his theory could be used "to engineer a legal outcome in or out of the courtroom, to reform a legal system, or even to design a community with little or no law at all." Such statements imply that law can be engineered, but "This can only be true if in some sense these variables cause law" (Greenberg 1983: 340). In preceding chapters, we have introduced issues/problems pertaining to causal evidence, and our arguments need not be repeated here. It will suffice to say that Black scarcely confronts the evidentiary problems, much less resolves them.

Whether or not you agree with Greenberg's assessment of the "causal" nature of assertions made in *The Behavior of Law*, claims for causality become explicit in *Moral Time*, where Black writes (2011: xii): "I have discovered that the fundamental cause of conflict is the movement of social time. And because conflict is a movement of social time, conflict causes still more conflict. Social time is moral time." Now things get interesting.

Black's formulation of the causal links among conflict, social time, and moral time brings to mind the analytic and synthetic distinction discussed in Chapter 4. If, as Black states (2011: xii), "The purpose of the theory of moral time is scientific: to explain conflict", one must ask: are Black's claims true or false by virtue of the very meaning of the constituent words or syntax?

The question leaves the reader wondering. If conflict is movement of social time and if social time is moral time, is Black's argument circular? At the very least there are doubts as to how Black's variables might be independently measured in any specified context. In addition, whatever the eventual measurement strategies for each variable, the asserted causal linkage between them will be, again for reasons set forth in preceding chapters, illusive in the extreme.

Predictive Accuracy

Setting aside the issues/problems in causal analysis, circular reasoning, and the analytic/synthetic distinction, what about the predictive accuracy of Black's theory? In that connection consider the systematic attempt by Gottfredson and Hindelang (1979a) to test a subset of the propositions in *The Behavior of Law*. They were particularly interested in the asserted link between stratification and the quantity of law.

As regards "quantity of law," Black asserts: "A complaint to a legal official, for example, is more law than no complaint, whether it is a call to the police, a visit to a regulatory agency, or a lawsuit. Each is an increase in the quantity of law" (1976: 3). When it comes to generalizations about the quantity of law, Black proposes, among others, "law varies directly with rank [e.g., income] … People with less wealth have less law. They are less likely to call upon the law in dealing with one another" (1976: 17). "In the case of a crime, for instance, a victim who is above the offender in rank is more likely to call the police than a victim whose rank is lower than the offender's" (Black 1976: 21).

These propositions are clearly stated; and in their tests of a portion of Black's framework Gottfredson and Hindelang (1979a: 5) drew data from the National Crime Survey (NCS), arguing that these data "are particularly well-suited to testing many of Black's generalizations because he explicitly frames much of his discussion in terms of criminal law examples, such as complaints to the police, arrests, and prosecutions." Gottfredson and Hindelang allege (1979a: 6) that "Contrary to Black's hypothesis … there is little systematic variability in rates of reporting to the police, as a function of income." Instead, the "seriousness of the offense" seemed to overwhelm any effect that might otherwise have been present.

This conclusion goes directly contrary to Black's claim (1976: 118) that the behavior of law can be explained "without regard to the … conduct of the deviant." As a modification of Black's generalizations, Gottfredson and Hindelang argued (1979a: 17): "To be consistent with the available data, an adequate theory of the behavior of criminal law must incorporate a proposition stating that the quantity of criminal law varies directly with the seriousness of the infraction."

So far so good. Conceptual problems are noted, assertions made, and tests conducted. Science at its best. Uncertainties emerge, however, in the subsequent exchange between Black (1979) and Gottfredson and Hindelang (1979b). Black claimed his colleagues were "misled by common sense" (1979: 18), noting in particular that Hindelang and Gottfredson failed to recognize that "to designate conduct of this kind as more 'serious' does not explain anything … but only begs the question of why it is handled as more serious" (1979: 25). In their rejoinder, Gottfredson and Hindelang (1979b: 30) pointed to inconsistencies in Black's definitions and reminded the reader: "In *The Behavior of Law*, Black states that his theory of law predicts the quantity of law at every state, from the initial definition of what is illegal, through calling the police, to arrest, incarceration, and release on parole." So Gottfredson and Hindelang maintained that Black's dismissal of their findings was based on inconsistent definitions of central concepts, thereby reducing both the testability of his ideas and greatly reducing the scope and range of generalizations in *The Behavior of Law*.

Lest it appear that the exchange between Black and Gottfredson/Hindelang is proof of no progress in sociology, that conclusion would be erroneous in ignoring two considerations. The first is that Gottfredson and Hindelang did not test all of Black's generalizations. For that matter, they did not so much as consider any dimension of predictive power other than accuracy, and that limitation is not just a matter of the terminology employed in Chapter 4. The crucial point is that there are few actual empirical theories about law in any literature, and of those known to the authors of this book none has the range and scope of Black's theory. So to the credit of Gottfredson and Hindelang, they did not claim to have disposed of all of Black's theory.

The second consideration is that Gottfredson and Hindelang did not compare the *apparent* predictive accuracy of Black's theory with that of some other theory. Yes, they do suggest that the "seriousness" of an event permits more accurate predictions of reactive characteristics than do Black's variables; but such a suggestion is a far cry from a theory, much less one that would remotely match the range and scope of Black's theory. Most important, all parties to the exchange should be chided for not simply revising Black's theory so as to include additional variables. As for Black's dismissal of "seriousness" because it is left unexplained, that is sheer nonsense. Apparently the dismissal reflects nothing more than Black's determination to "purify" sociology by driving out all psychological notions. What a high price to pay! Contemplate someone condemning Newton's gravitational theory because it does not explain variation in mass!

Should a critic protest that simply adding variables to a theory opens the door to mindless and endless reformulations, he/she will have ignored "parsimony" as a predictive power dimension. But so did Gottfredson and Hindelang.

Finally, for present purposes, Black's theory clearly illustrates the importance of the search for uniformities. Whatever the shortcomings of the theory, it is an obvious product of an extensive search for uniformities. To be sure, someone with Black's talent must translate the identified uniformities into a theory, but that does not make the search for uniformities less important.

Seeking the Principles of Sociology

It is safe to say that the basic principles of any science can be elusive. Even so, attempts to formulate those principles are preferable to what humanists and "phrase makers" in the social sciences regard as profound generalizations. Three examples:

Lord Acton: Power tends to corrupt and absolute power corrupts absolutely.
Leo Tolstoy: All happy families are alike; each unhappy family is unhappy in its own way.
Marshall McLuhan: The Medium is the message.

Although these statements provoke smiles among some and somber nods of agreement among others, they rarely result in anything approaching science as here conceived. In the case of sociology, yearning for the profound discourages searching for uniformities in research findings. That search is discouraged because possible uniformities are likely to be dismissed as "trivial" or all too "mundane." The really crucial point is that "profound statements" are commonly of no scientific value whatever.

In Lord Acton's case "power" remains one of the fuzziest notions in human thought, even though sociologists use the word with abandon, taking its meaning for granted, as real and tangible as a fireplug. Even when treated with more conceptual precision as a theory-related construct (e.g. Emerson 1962; Cook and Emerson 1978; Cook and Yamagishi 1992), there are diverse conceptual problems unless power is defined as the perceived capacity for effective control (for elaboration, see Gibbs 2008: 44–49 but also Roscigno 2011). We leave Tolstoy and McLuhan for another day and turn to an example of a serious search for the basic principles of sociology, a search that goes back at least to Herbert Spencer's *The Principles of Sociology* (commencing 1874).

Jonathan Turner's Monumental Effort

We now turn with critical admiration to Jonathan Turner's effort (2010, two volumes, with a promised third) to identify sociology's principles. Unfortunately, Turner seeks a renewal of "grand theorizing" but unlike sociologists who are content to warm their hands over ancestral bones, he formulated twenty-three theoretical principles on "macrodynamics" in his first volume, the exclusive concern here.

Although Turner identifies his formulations as a theory, in keeping with the tradition of grand theory he left the meaning of "a theory" obscure. The closest he comes to an explicit definition of a theoretical principle comes early in his discussion (2010: 7): "The ultimate criterion for good theoretical principle is this: does it denote the dynamics of some property of the universe that is always present when humans interact and organize?" We leave the vagaries of this explication for the reader to contemplate. We say with confidence, however, that empirical applicability was apparently the least of Turner's concerns when he formulated principles. They assert relations among hundreds of variables and Turner scarcely considered any measurement procedures.[5] And only rarely does he cite research findings in connection with a *component* of one of the twenty-three principles.

We emphasize "component" because in Turner's scheme each principle comprises a large number of independent assertions, independent in the sense that the outcome of a test of one assertion would have no necessary bearing on the others. Indeed, Turner only occasionally uses the words "test" or "tests," and it is difficult to see how any of his principles could be tested, all the more given Turner's apparent indifference to testability in his critique of "functionalism" (2010: 22–23). Though otherwise very informative, there is no reference whatever in that critique to test, tests, or testability, just as in the case of Kingsley Davis's critique (1959) of functionalism.

To be sure, all of Turner's principles can be construed as generalizations (hence, possibly assertions of uniformities), even though many of them have no unit term. As noted previously, such omission is a serious shortcoming, for it taxes credulity to assume that any sociological generalization holds for all possible entities (e.g., countries, tribes, cities, families, bands, to name just a few), let alone holds even approximately the same. Nevertheless, other than making occasional references to "society," Turner appears completely indifferent to the very notion of a unit term.

Consider the sixth major component of the first principle (2010: 624–625): "1. The size of the population is: F. A positive function of the degree of institutional differentiation and the formation of distinctive cultures to regulate relations among actors within and between institutional domains … G." Observe there is no stipulation of a unit term, meaning no indication as to variation in population size among what kinds of populations. But can you imagine any correlate of population size that holds regardless of the kinds of populations that are compared (e.g., residents of countries as well as residents of villages).

Setting unit specification aside, should you, the reader, be a sociologist and undertake a test of Turner's major component "*F*," how would you measure the constituent variables?[6] We find it difficult to even imagine an empirically applicable measurement procedure for any of the constituent variables other than population size, and even here a specification of the entities in question is needed. For what it is worth, both authors of this book truly wonder what Turner means by "institutional differentiation," and we would balk at even attempting to measure "the formation of distinctive cultures to regulate actors." Should this obsession with measurement irritate, note that Turner himself speaks of "degree." Assuming normal understandings of what that word means, it is safe to say that Turner created the measurement problem.

Suppose Turner grants that few of his principles are systematically testable. Even so, he could argue that outcomes of tests of those that are testable would shed light on the untestable. Indeed, that would be the case if testable principles could be *deduced* from the untestable and thereby provide evidence for or against the untestable. Unfortunately, Turner did not state his principles in such a manner. To the contrary, he writes (2010: 6) "… I reject the presumption that nomothetic theory must be axiomatic and reveal deductive rigor."

At this point, the reader might be thinking: you fellows are really criticizing Turner because he did not use formal theory construction. To the contrary, a close reading of Turner's work reveals marks (or "stigma," depending on your perspective) of a formally stated theory. For one thing, his relational terms are exact equivalents of those prescribed in Chapter 6. For another, his emphasis on "lagged function" clearly reflects an all too rare awareness by sociological theorists of the difference between spatial and temporal relations. As for Turner's systematic and explicit listing of his principles apart from definitions and other kinds of statements (what we would call the "intrinsic part" of a theory), that feature alone is a departure from conventional use of a natural language to state a theory.

So it is that despite our carping, we truly envy and admire Turner's work, in recognition of its enormous scope but also the product of a fertile imagination with obvious dedication to the importance of sociological theory. Above all, it is the promise inherent in Turner's principles that commands our respect the most. If Turner or his paladins should ever express his principles in light of some mode of formal theory construction (MFTC), it would deserve fervid acclamation. The most appropriate MFTC remains an open question, but to clarify, we will describe what could be done in light of our proposed MFTC (Chapter 6).

Turner's principles would be restated as generalizations with each having what is considered the most appropriate unit and characteristic terms; and those generalizations that comprise only constructs would be identified as axioms. Then suppose that several of Turner's variables are defined such as to justify identifying them as concepts. Those identifications would set the scene for linking the concepts with the constructs (not necessarily all), with those links asserted as postulates. The axioms and postulates would be sufficient for deducing all manner of theorems. Of course, the final step would be epistemic statements that link the concepts with referentials, each denoting a measurement procedure and kinds of requisite data, in light of which the theorems would become *testable*. Regardless of the outcomes, reports of tests would themselves be an occasion for celebration by those sociologists who think of their field as a genuine science.

Our criticisms of Turner's work in no way contradict the argument that a search for uniformities could lead sociology out of its wilderness. But our criticisms underscore a related point, that the outcome of a search depends entirely on the way the detected uniformities are expressed.

Formal Theory Construction, Uniformities, and Propositions

Whether formally stated or not, a theory's theorems qualify as assertions of uniformities. The same may be said of propositions standing apart from any particular theory. Indeed, stand-alone uniformities expressed as propositions may stimulate efforts at theory construction. As defined previously (Chapter 6), the constituent characteristic terms in a proposition are concepts, and the empirical applicability of concepts enhances the testability of propositions. But, to repeat, propositions may stimulate theories.

The attention devoted to uniformities and propositions will not be appreciated unless it is recognized that the "go theorize" admonition is sterile. Granted that there can be no logic of discovery and that a theory may be the product of brilliant intuition, impressive sociological theories are much more likely if formulated with a view to deducing what commenced as the detection of uniformities. Stating our argument most extremely, the road to theories is more likely to be paved by the pursuit of uniformities than by anything else, *including* adoption of some MFTC. Nevertheless, whatever the path to a theory—a search for uniformities or brilliant intuition—its predictive power will be enhanced if stated in accordance with some formal mode.

More on the Strategic Role of Formal Theory Construction

Assuming that sociologists commence formulating explicit propositions as summaries of detected uniformities in the findings of descriptive or exploratory studies, there would be a crucial question. It bears on one of this book's major questions: is formal theory construction (FTC) absolutely essential to convert the propositions into theories? The answer is negative, and it is refreshing to admit that the prescribed mode of FTC (Chapters 6 and 7) is no exception. Nonetheless, we persist and suggest for reasons given subsequently that use of the mode's terminology in stating propositions would be most desirable.

The Explicit and Consistent Use of Unit Terms

Assume a statement in some sociology journal something like this: The findings clearly indicate a *negative* association between gender discrimination in hiring practices and the position vacancy rate. While informative, there is a glaring unanswered question: associated among what? The possibilities are endless: among Nebraska public schools, 2013; for General Motors, over the period 2000–2015; for all chemistry departments in public universities in the U.S., between 1990–2010; among private hospitals in New York City, 2015. Clearly, these specifications involve unit terms, and the association in question could be much greater for some than others. The crucial point: without an explicit unit term, a summary of findings cannot be a coherent proposition.

This insistence on explicit unit terms (each denoting some kind of entity) is not a concealed attempt to impose some orthodoxy. Unit terms come in many varieties, including all-encompassing terms like "organizations," all designed to formulate a theory not limited in space or time. The general point is that the use of a MFTC does not dictate the content of the theory, commencing with the choice of a unit term *or* unit terms. Even "among individuals" is a possibility, and a unit term can be such as to designate multiple kinds of entities (e.g., "among human populations"). Indeed, if the association in question holds even more or less for multiple kinds of entities, that consistency could have considerable theoretical significance, certainly greater confidence in the asserted association. Whatever, for various reasons, it is imperative that any assertion of an association include an explicit unit term.

The "Characteristic Terms" in Propositions

Recall that a proposition is defined in Chapter 6 as an assertion in which the constituent characteristic terms are concepts. Then also recall that concepts are defined as being empirically applicable, which translates as "measurable" when the concepts are linked to referentials through epistemic statements. What is the importance of that definition of a proposition? Even if it guides a literature search to detect uniformities, there would be no assurance of the outcome being a theory, not even if several detected uniformities are expressed as propositions. After all, the formulation of a theory requires creativity and judgment, which is to say that there is no mechanical process or rule through which a list of propositions (whatever the source) will necessarily result in a theory. Nevertheless, if such a list of propositions does somehow inspire a theory, testability is virtually assured.

To avoid confusion and alleviate doubts, it is desirable to state the minimum requirements we have in mind for a theory generated in part by a search for uniformities and their renditions as propositions. Such a theory must comprise *at least* two propositions, with illustrative instances and the implied theorem, deduced by the sign rule, stated abstractly as:

Proposition I: Among *U*s, the greater *W*, the greater *X*.
Proposition II: Among *U*s, the greater *X*, the less *Y*.
Theorem: Among *U*s, the greater *W*, the less *Y*.

Assume that *W*, *X*, and *Y* are concepts and *U* is some conventional unit term in sociology (e.g., cities in the same country, age groups in any city, families with same-sex partners, etc.). Hence, theorems would be derived from the propositions not primarily with a view to *indirect tests* of the propositions but rather to identify some relation (the theorem) heretofore unrecognized.[7]

Standard Relational Terms

Even a truncated survey of the sociological literature will reveal that sociologists use diverse terms to identify generalizations, such as: hypothesis, proposition, or simply something like "summary of findings." But that variety is negligible compared to the diversity of relational terms embedded in those generalizations. Here is a sample of ten among the virtually uncountable: causes, is necessary for, is sufficient for, produces, leads to, results in, influences, shapes, determines, and limits.

The sheer variety of relational terms would *not* be a truly crippling problem if one detects uniformities solely in the hope that they will somehow give rise to ideas, as opposed to a well formulated theory. The situation is quite different, however, if the detected uniformities are to be expressed as axioms, postulates, or propositions in what would be a formally stated theory.[8] If that is the goal, it is imperative that the relational terms be standardized, as they would be if translated into those prescribed in Chapter 6. But alternatives could be considered, and the same may be said of the use of the sign rule in deducing theorems. However, it would be a monumental delusion to think that testable conclusions can be deduced from causal premises without translating the relational terms into space-time relations.

Bottom Lines

Well, we come to the end of what has been a long road. An attempt to summarize all that we have covered may be futile, but we will take a stab at it.

To be considered scientists in the full sense of that term, sociologists must set as their top priority the formulation of propositions stemming from detection of possible uniformities in what are sociological phenomena by one criterion or another. Of course, as has always been the case, research projects could be launched with a view to detecting uniformities, but resources for original studies (as opposed to literature searches) will be and have always been limited. For that matter, the existing sociological literature has enormous potential for detecting uniformities, some of which may be explicit. The major immediate problem is the diverse terminology employed in reporting research findings, a diversity that can be overcome *only* by diminishing the reliance on natural language conventions.

At no point have we proposed some delimitation of sociology's subject matter. Moreover, we gleefully grant that the search for uniformities and expressing them as propositions would further the scientific standing of sociology more than would even an all-out adoption of some MFTC. But that standing will not be furthered unless sociologists cease their promotion of ideological causes and more intensely engage in descriptive exploratory research with a view to detecting uniformities and expressing them as propositions.

Finally, lest our arguments be dismissed as conservative or reactionary, as far as we are concerned, there is no scientific limit on the subject matter of any science or the kind of questions that scientists can pursue. Nevertheless, whatever the question pursued, acceptance/rejection of the answer should be governed by the supreme criterion of relative predictive power.

Notes

1 See: http://plato.stanford.edu/entries/double-effect/
2 See also: http://carnegie.org/fileadmin/Media/Publications/fall_04americandilemma.pdf/
3 While Moynihan's report had great influence on President Lyndon Johnson's "War on Poverty," it was also greeted with derision and charges of "blaming the victim" by many critics (e.g., Ryan 1971).
4 See also: http://www.baylorisr.org/2012/06/20/a-social-scientific-response-to-the-regnerus-controversy/ and https://familyinequality.wordpress.com/2012/06/29/200-researchers-respond-to-regnerus-paper/
5 The terminology in Chapter 6 provides a neat way to succinctly summarize Turner's "grand theorizing." In keeping with his predecessors, Turner's language is a study in constructs (e.g., "institutional differentiation") rather than concepts (e.g., "population size") with no obvious concern as regards epistemic statements.
6 Lest we, the authors, be accused of selecting the most complicated of Turner's principles, we urge you to examine Principle 6 (2010: 328–330). Its statement runs over a page, a total of fifty-three lines, with twenty major or minor components.
7 Of course, admirers of "grand theorizers," unencumbered by tests or testability, commencing with A. Comte and fortunately ending with T. Parsons, might see such a theoretical framework as trivial and perhaps not a theory at all. We would reply with confidence that the sterility of that tradition is all too obvious. Jonathan Turner (2010) would likely take exception to such a reply, but that is all the more a reason for careful consideration of his notable revival of "grand theorizing."
8 Note especially that a detected uniformity need not be identified as a proposition. Such identification has been stressed previously but only with a view to testability.

APPENDIX I: ELEVEN DEFINITIONS OF SCIENCE

Definition I: "The modern restricted sense of a branch of learning based on observation and tested truths, arranged in an orderly system, is first recorded in English in Israel Watt's *Logic* (1725) …" (Barnhart 1988: 968).

Definition II: "The definition of the term 'science' is problematic. Science can refer to a changing body of shared knowledge about nature or to the methods used to obtain that knowledge; in that form, science has existed for millennia" (Hagstrom 2000: 2455).

Definition III: "The term science … means 'knowledge.' In philosophy it refers strictly to proven ideas, to the exclusion of hypotheses or speculations. Until the twentieth century, proof remained mysterious, but what it achieves has been clear since antiquity: certainty, truth unshakable by criticism or doubt" (Agassi 2008: 357).

Definition IV: "1. The observation, identification, description, experimental investigation, and theoretical explanation of natural phenomena. 2. Such activity restricted to a class of natural phenomena. 3. Such activity applied to any class of phenomena. 4. Any methodological activity, discipline, or study. 5. Any activity that appears to require study and method. 6. Knowledge; especially, knowledge gained through experience" (Morris 1969: 1162).

Definition V: "**1a**. The observation, identification, description, experimental investigation, and theoretical explanation of phenomena: *new advances in science and technology*. **b**. Such activities restricted to a class of natural phenomena: *the science of astronomy*. **2**. A systematic method or body of knowledge in a given area: *the science of marketing*. **3**. *Archaic* Knowledge, especially that gained through experience" (Pickett 2011: 1570).

Definition VI: "the intellectual and practical activity encompassing the systematic study of the structure and behavior of the physical and natural world through observation and experiment: *the world of science and technology*. ■ a particular area of this: *veterinary science* | *the agricultural sciences*. ■ a systematically organized body of knowledge on a particular subject: *the science of criminology*. ■ *archaic* knowledge of any kind" (Stevenson and Linberg 2010: 1564).

Definition VII: "1: the state of knowing: knowledge as distinguished from ignorance or misunderstanding **2 a:** a department of systematized knowledge as an object of study <the ~ of theology> **b:** something (as a sport or technique) that may be studied or learned like systematized knowledge <have it down to a ~> **3 a:** knowledge or a system of knowledge covering general truths or the operation of general laws esp. as obtained and tested through scientific method **b:** such knowledge or such a system of knowledge concerned with the physical world and its phenomena: *natural science* **4:** a system or method reconciling practical ends with scientific laws <cooking is both a ~ and an art> **5** *cap*: ***CHRISTIAN SCIENCE***" (Mish 2003: 1112).

Definition VIII: "knowledge ascertained by observation and experiment, critically tested, systematized and brought under general principles, *esp* in relation to the physical world; a department or branch of such knowledge or study; knowledge (*archaic*); a skilled craft (*obs*); trained skill, *esp* in boxing (*now usu joc*)" (Chambers Harrap Publishers 2011: 1396).

Definition IX: "The ordered arrangement of ascertained knowledge, including the methods by which such knowledge is extended and the criteria by which its truth is tested" (Walker 1999: 1021).

Definition X: "**1. a.** The state or fact of knowing; knowledge or cognizance *of* something specified or implied; also, with wider reference, knowledge (more or less extensive) as a personal attribute … **2. a.** knowledge acquired by study; acquaintance with or mastery of any department of learning … **3. a.** A particular branch of knowledge or study; a recognized department of learning … **4. a.** In a more restricted sense: A branch of study which is concerned either with a connected body of demonstrated truths or with observed facts systematically classified and more or less colligated by being brought under general laws, and which includes trustworthy methods for the discovery of new truth within its own domain … **5. a.** The kind of knowledge or of intellectual activity of which the various 'sciences' are examples" (Simpson and Weiner 1989, Vol. XIV: 649).

Definition XI: "The document was an amicus curiae brief submitted to the court on behalf of seventy-two Nobel laureates in science, seventeen state academies of science, and seven other scientific organizations. The amicus brief begins by offering a general definition: 'Science is devoted to formulating and testing naturalistic explanations for natural phenomena. It is a process for systematically collecting and recording data about the physical world, then categorizing and studying the collected data in an effort to infer the principles of nature that best explain the observed phenomena'" (Shermer 2005: 670–671).

APPENDIX II: EIGHT DEFINITIONS OF THEORY

Definition 1: "The sense of the principles or methods of a science or art rather than its practice is first recorded in Robert Cawdrey's *A Table Alphabeticall* (1613). The sense of an explanation based on observation and reasoning is first recorded in 1638" (Barnhart 1988: 1132).

Definition 2: "**1. a.** Systematically organized knowledge applicable in a relatively wide variety of circumstances; especially, a system of assumptions, accepted principles, and rules of procedure devised to analyze, predict, or otherwise explain the nature or behavior of a specified set of phenomena. **b.** Such knowledge or such a system distinguished from experiment or practice. **2.** Abstract reasoning; speculation. **3.** Broadly, hypothesis or supposition" (Morris 1969: 1335).

Definition 3: "**1.** A set of statements or principles devised to explain a group of facts or phenomena, especially one that has been repeatedly tested or is widely accepted and can be used to make predictions about natural phenomena. **2.** The branch of a science or art consisting of its explanatory statements, accepted principles, and methods of analysis, as opposed to practice: *a fine musician who had never studied theory*. **3.** A set of theorems that constitute a systematic view of a branch of mathematics. **4.** Abstract reasoning; speculation: *a decision based on experience rather than theory*. **5.** A belief or principle that guides action or assists comprehension or judgments: *staked out the house on the theory that criminals usually return to the scene of the crime*. **6.** An assumption based on limited information or knowledge; a conjecture" (Pickett 2011: 1805).

Definition 4: "a supposition or a system of ideas intended to explain something, esp. one based on general principles independent of the thing to be explained: *Darwin's theory of evolution*. ■ a set of principles on which the practice of an activity is based: *a theory of education | music theory*. ■ an idea used to account for a situation or justify a course of action: *my theory would be that the place has been seriously mismanaged*. ■ *Mathematics* a collection of propositions to illustrate the principles of a subject" (Stevenson and Lindberg 2010: 1799).

Definition 5: "**1.** the analysis of a set of facts in their relation to one another **2.** abstract thought: *SPECULATION* **3.** the general or abstract principles of a body of fact, a science, or an art <*music* ~> **4a.** a belief, policy, or procedure proposed or followed as the basis of action <her method is based on the ~ that all children want to learn> **b:** an ideal or hypothetical set of facts, principles, or circumstances – often used in the phrase *in theory* <in ~ we have always

advocated freedom for all> **5.** a plausible or scientifically acceptable general principle or body of principles offered to explain phenomena <the wave ~ of light> **6a.** a hypothesis assumed for the sake of argument or investigation **b.** an unproved assumption: ***CONJECTURE*** **c.** a body of theorems presenting a concise systematic view of a subject <~ of equations>" (Mish 2003: 1296).

Definition 6: "an explanation or system of anything; an exposition of the abstract principles of a science or art; an idea or explanation that has not yet been proved, a conjecture; speculation as opposed to practice; an ideal, hypothetical or abstract situation, *esp.* in the phrase *in theory*; ideal, hypothetical, or abstract reasoning" (Chambers Harrap Publishers 2011: 1618).

Definition 7: "**1.** A sight, a spectacle. *Obs. rare* … **2.** Mental view, contemplation. *Obs* … **3.** A conception or mental scheme of something to be done, or of the method of doing it; a systematic statement of rules or principles to be followed … **4.a.** A scheme or system of ideas or statements held as an explanation or account of a group of facts or phenomena; a hypothesis that has been confirmed or accepted as established by observation or experiment, and is propounded or accepted as accounting for the known facts; a statement of what are held to be the general laws, principles, or causes of something known or observed … **b.** That department of an art or technical subject which consists in the knowledge or statement of the facts on which it depends, or of its principles or methods, as distinguished from the *practice* of it … **c.** A systematic statement of the general principles or laws of some branch of mathematics; a set of theorems forming a connected system; as **the theory of equations, of functions, of numbers, of probabilities** … **5.** In the abstract (without article): Systematic conception or statement of the principles of something; abstract knowledge, or the formulation of it; often used as implying more or less unsupported hypothesis … distinguished from opposed to *practice* … **6.** In loose or general sense: A hypothesis proposed as an explanation; hence, a mere hypothesis, speculation, conjecture; an idea or set of ideas about something; an individual view or nation" (Simpson and Weiner 1989, Vol. XVII: 902).

Definition 8: "The notion of a theory is controversial in social science. A single and simple conception of theory is unlikely to apply across all fields, from mathematical economics to cultural anthropology. Still, construing *theory* broadly as any attempt to systematize and explain certain phenomena, it is clear that theories play a central role in social science" (Ellis and Swoyer 2008: 343).

REFERENCES

Abend, Gabriel. 2008. "The Meaning of 'Theory'." *Sociological Theory*, 26: 173–199.

Agassi, Joseph. 2008. "Science," pp. 357–360 in William A., Darity, Jr., ed., *International Encyclopedia of Social Sciences*, 2nd ed., Vol. 7. Detroit, MI: Gale, Cengage Learning.

Alessio, John C. 1982. "Axiomatic Theories and Syllogistic Form." *International Review of Modern Sociology*, 12: 111–134.

Anderson, Benedict. 2006. *Imagined Communities* (Revised). New York: Verso.

Anderson, Elijah. 1990. *Street Wise: Race, Class, and Change in an Urban Community*. Chicago, IL: University of Chicago Press.

Annals of the American Academy of Political and Social Science. 2009. *The Moynihan Report Revisited: Lessons and Reflections after Four Decades*. Thousand Oaks, CA: Sage Publications.

Arendt, Hannah. 1963. *Eichmann in Jerusalem: A Report on the Banality of Evil*. New York: Viking Press.

Auyero, Javier. 2009. *Flammable: Environmental Suffering in an Argentine Shantytown*. New York: Oxford University Press.

Baesens, Bart. 2014. *Analytics in a Big Data World: The Essential Guide to Data Science and its Applications*. Hoboken, NJ: John Wiley & Sons.

Barnes, Simon. 2014. *Ten Million Aliens: A Journey through the Entire Alien Kingdom*. New York: Marble Arch Press.

Barnhart, Robert K., ed. 1988. *Chambers Dictionary of Etymology*. New York: Chambers Harrap and H.W. Wilson.

Beller, Emily and Michael Hout. 2006. "Intergenerational Social Mobility: The United States in Comparative Perspective." *The Future of Children*, 16: 19–36.

Bengtson, Vern L. and Robert E.L. Roberts. 1991. "Intergenerational Solidarity in Aging Families: An Example of Formal Theory Construction." *Journal of Marriage and the Family*, 53: 856–870.

Berger, Peter L. and Thomas Luckman. 1966. *The Social Construction of Reality: A Treatise in the Sociology of Knowledge*. New York: Doubleday.

Bernard, Thomas J. 2002. "Donald Black's Influence in Criminology." *Contemporary Sociology*, 31: 650–652.

Bernstein, Richard J. 2008. "Are Arendt's Reflections on Evil Still Relevant?" *Review of Politics*, 70: 64–76.

Bestor, Theodore C. 2004. *Tsukiji: The Fish Market at the Center of the World*. Berkeley: University of California Press.

Bittman, Michael and Jocelyn Pixley. 2004. "Equality and Gender, Divided versus Multiple Subjectivity: A Response to Allon J. Uhlmann's 'The Sociology of Subjectivity, and the Subjectivity of Sociologists.'" *British Journal of Sociology*, 55: 433–438.

Black, Donald. 1976. *The Behavior of Law*. New York: Academic Press.

Black, Donald. 1979. "Common Sense in the Sociology of Law." *American Sociological Review*, 44: 18–27.

Black, Donald. 1995. "The Epistemology of Pure Sociology." *Law & Social Inquiry*, 20: 829–870.

Black, Donald. 2011. *Moral Time*. New York: Oxford University Press.

Black, Edwin. 2003. *War against the Weak: Eugenics and America's Campaign to Create a Master Race*. New York: Basic Books.

Black, Max, ed. 1961. *The Social Theories of Talcott Parsons*. Englewood Cliffs, NJ: Prentice Hall.

Blalock, H.M., Jr. 1964. *Causal Inferences in Non-Experimental Research*. Chapel Hill: University of North Carolina Press.

Blalock, H.M., Jr. 1967. *Toward a Theory of Minority-Group Relations*. New York: Wiley.

Blumer, Herbert. 1969. *Symbolic Interactionism*. Englewood Cliffs, NJ: Prentice Hall.

Bonevac, Daniel. 1990. *The Art and Science of Logic*. Mountain View, CA: Mayfield Publishing.

Bonjean, Charles M., Richard J. Hill, and S. Dale McLemore. 1967. *Sociological Measurement: An Inventory of Scales and Indices*. San Francisco, CA: Chandler.

Bothamley, Jennifer. 2002. *Dictionary of Theories*. Detroit, MI: Visible Ink Press.

Bowen, William G. and Derek Bok. 1998. *The Shape of the River: Long-Term Consequences of Considering Race in College and University Admissions*. Princeton, NJ: Princeton University Press.

Broom, Leonard, Philip Selznick, and Dorothy H. Broom. 1981. *Sociology*, 7th ed. New York: Harper & Row.

Buss, David M. 2004. *Evolutionary Psychology: The New Science of the Mind*, 5th ed. Boston, MA: Pearson.

Caldwell, John C. 2010. *Demographic Transition Theory*. Dordrecht: Springer.

Camerer, Colin F., George Loewenstein, and Drazen Prelec. 2004. "Neuroeconomics: Why Economics Needs Brains." *The Scandinavian Journal of Economics*, 106: 555–579.

Carey, Nessa. 2012. *The Epigenetics Revolution: How Modern Biology Is Rewriting our Understanding of Genetics, Disease and Inheritance*. New York: Columbia University Press.

Chalmers, Alan. 1990. *Science and Its Fabrication*. Minneapolis: University of Minnesota Press.

Chambers Harrap Publishers. 2011. *The Chambers Dictionary*, 12th ed. London: Chambers Harrap.

Choi, Bernard C.K. and Anita W.P. Pak. 2005. "A Catalog of Biases in Questionnaires." *Preventing Chronic Disease*, 2: http://www.cdc.gov/pcd/issues/2005/jan/04_0050.htm.

Clark, Rob. 2013. "Convergence in National Income Distributions." *Social Forces*, 92: 413–436.

Clayton, Obie, ed. 1996. *An American Dilemma Revisited: Race Relations in a Changing World*. New York: Russell Sage Foundation.

Collins, Randall. 2012. "C-Escalation and D-Escalation: A Theory of the Time-Dynamics on Conflict." *American Sociological Review*, 77: 1–20.

Cook, Karen S. and Robert M. Emerson. 1978. "Power, Equity, and Commitment in Exchange Networks." *American Sociological Review*, 43: 721–739.

Cook, Karen S. and Toshio Yamagishi. 1992. "Power in Exchange Networks: A Power-Dependence Formulation." *Social Networks*, 14: 245–265.

Coser, Lewis. 1956. *The Functions of Social Conflict*. New York: Free Press.

Costner, Herb. 1965. "Criteria for Measures of Association." *American Sociological Review*, 30: 341–353.

Cromer, Alan. 1993. *Uncommon Sense*. New York: Oxford University Press.

Darwin, Charles. 1876. *Autobiography of Charles Darwin*, edited by N. Barlow. London: Collins.

Darwin, Charles. 1996 [1859]. *The Origin of Species*. New York: Oxford University Press.

Davis, Kingsley. 1945. "The World Demographic Transition." *Annals of the American Academy of Political and Social Science*, 237: 1–11.

Davis, Kingsley. 1959. "The Myth of Functional Analysis as a Special Method in Sociology and Anthropology." *American Sociological Review*, 24: 757–772.

Dennett, Daniel C. 2013. *Intuition Pumps and Other Tools for Thinking*. New York: W. W. Norton.

Desmond, Matthew. 2007. *On the Fireline: Living and Dying with Wildland Firefighters*. Chicago, IL: University of Chicago Press.

Deutsch, David. 2011. *The Beginning of Infinity*. New York: Viking, Penguin.

Douglas, Jack D. 1967. *The Social Meanings of Suicide*. Princeton, NJ: Princeton University Press.

Duncan, Otis D., ed. 1964. *William Ogburn on Culture and Social Change*. Chicago, IL: University of Chicago Press.

Duncan, Otis D. 1966. "Path Analysis: Sociological Examples." *American Journal of Sociology*, 72: 1–16.

Durkheim, Emile. 1949 [1893]. *The Division of Labor in Society*, translated by George Simpson. New York: Free Press.

Durkheim, Emile. 1951 [1897]. *Suicide*. New York: Free Press.

Ehrlich, Robert. 2003. *Eight Preposterous Propositions*. Princeton, NJ: Princeton University Press.

Einstein, Albert. 1918. "Principles of Research." Speech given for Max Planck's 60th Birthday. http://speakola.com/ideas/albert-einstein-principles-of-research-1918.

Einstein, Albert. 1921. Address to Prussian Academy of Sciences, 1921. http://www-groups.dcs.st-and.ac.uk/history/Extras/Einstein_geometry.html.

Ekland-Olson, Sheldon. 2012. *Who Lives, Who Dies, Who Decides?* New York: Routledge.

Ellis, Stephen and Chris Swoyer. 2008. "Theory," pp. 343–345 in William A. Darity Jr., ed. *International Encyclopedia of the Social Sciences*, 2nd ed., Vol. 8. Detroit, MI: Gale, Cengage Learning.

Emerson, Robert M. 1962. "Power-Dependence Relations." *American Sociological Review*, 27: 31–40.

Erickson, Kai. 1976. *Everything in its Path: Destruction of Community in the Buffalo Creek Flood*. New York: Simon & Schuster.

Fararo, Thomas J. 1997. "Reflections on Mathematical Sociology." *Sociological Forum*, 12: 73–101.

Feagin, Joe. 2001. "Social Justice and Sociology in the 21st Century." *American Sociological Review*, 66: 1–20.

Feyerabend, P. 1975. *Against Method*. London: Verso.

Foner, Eric. 1988. *Reconstruction: America's Unfinished Revolution, 1863–1877*. New York: Harper & Row.

Foreman, John W. 2013. *Data Smart: Using Data Science to Transform Information into Insight*. Indianapolis, IN: Wiley.

Frankfort-Nachmias, Chava and Anna Leon-Guerrero. 2014. *Social Statistics for a Diverse Society*, 6th ed. Los Angeles, CA: Sage Publications.

Fraser, Steven, ed. 1995. *The Bell Curve Wars: Race Intelligence, and the Future of America*. New York: Basic Books.

Garfinkel, Harold. 1967. *Studies in Ethnomethodology*. Englewood Cliffs, NJ: Prentice Hall.

Gazzaniga, Michael S. 2005. *The Ethical Brain: The Science of Our Moral Dilemmas*. New York: Dana Press.

Gergen, Kenneth. 1994. *Toward Transformation of Social Knowledge*, 2nd ed. London: Sage Publications.

Gergen, Kenneth. 1999. *An Invitation to Social Construction*. London: Sage Publications.

Gibbs, Jack P. 1975. *Crime, Punishment, and Deterrence*. New York: Elsevier.

Gibbs, Jack P. 1989. *Control: Sociology's Central Notion*. Urbana: University of Illinois Press.

Gibbs, Jack P. 1994. *A Theory About Control*. Boulder, CO: Westview Press.

Gibbs, Jack P. 2001. "Deviant Cases in Tests of the Status Integration Theory." *Sociological Theory*, 19: 271–291.

Gibbs, Jack P. 2003. "A Formal Restatement of Durkheim's 'Division of Labor Theory'." *Sociological Theory*, 21: 103–127.

Gibbs, Jack P. 2008. *Colossal Control Failures*. Boulder, CO: Paradigm.

Gibbs, Jack P. and Walter T. Martin. 1964. *Status Integration and Suicide: A Sociological Study*. Eugene: University of Oregon Books.

Glynn, Ian. 2010. *Elegance in Science*. New York: Oxford University Press.

Goffman, Alice. 2014. *On the Run: Fugitive Life in an American City*. Chicago, IL: University of Chicago Press.

Goffman, Erving. 1961. *Asylums: Essays on the Social Situation of Mental Patients and Other Inmates*. New York: Anchor Books.

Goffman, Erving. 1963. *Stigma: Notes on the Management of Spoiled Identity*. New York: Simon & Schuster.

Goldthorpe, John H. 2001. "Causation, Statistics, and Sociology." *European Sociological Review*, 17: 1–20.

Gorham, Geoffrey. 2009. *Philosophy of Science*. Oxford: One World.

Gottfredson, Michael R. and Michael J. Hindelang. 1979a. "A Study of the Behavior of Law." *American Sociological Review*, 44: 3–18.

Gottfredson, Michael R. and Michael J. Hindelang. 1979b. "Theory and Research in the Sociology of Law." *American Sociological Review*, 44: 27–37.

Gray, David E. 1993. "Perceptions of Stigma: The Parents of Autistic Children." *Sociology of Health and Illness*, 15: 102–120.

Greenberg, David F. 1983. "Donald Black's Sociology of Law: A Critique." *Law & Society Review*, 17: 337–368.

Greene, Joshua. 2014. *Moral Tribes: Emotion, Reason, and the Gap between Us and Them*. New York: Penguin Books.

Hagstrom, Warren. 2000. "Science," pp. 2455–2463 in Edgar F. Borgatta and Rhonda J.V. Montgomery, eds., *Encyclopedia of Sociology*, 2nd ed., Vol. 4. New York: Macmillan (Gale Group imprint).

Halfpenny, Peter. 1982. *Positivism and Sociology*. London: Allen & Unwin.

Hammond, Phillip E., ed. 1967. *Sociologists at Work*. Garden City, NY: Doubleday.

Haney, Craig, Curtis Banks, and Philip Zimbardo. 1973. "A Study of Prisoners and Guards in a Simulated Prison." *Naval Research Reviews* (Online Publication).

Harrell, Frank E., Jr. 2001. *Regression Modeling Strategies: With Applications to Linear Models, Logistic Regression, and Survival Analysis*. New York: Springer Verlag.

Hedberg, Katrina, David Hopkins, Richard Leman, and Melvin Kohn. 2009. "The 10-Year Experience of Oregon's Death with Dignity Act: 1998–2007." *Journal of Clinical Ethics*, 20: 124–132.

Hedström, Peter and Petri Ylikoski. 2010. "Causal Mechanisms in the Social Sciences." *Annual Review of Sociology*, 36: 49–67.

Heidtman, Joanna, Kinga Wysienska, and Jacek Szmatka. 2000. "Positivism and Types of Theories in Sociology." *Sociological Focus*, 33: 1–26.

Hembroff, Larry A. 1987. "The Seriousness of Acts and Social Contexts: A Test of Black's Theory of the Behavior of Law." *American Journal of Sociology*, 93: 322–347.

Hempel, Carl G. and Paul Oppenheim. 1948. "Studies in the Logic of Explanation." *Philosophy of Science*, 15: 135–175.

Homans, George C. 1950. *The Human Group*. New York: Harcourt, Brace & World.

Homans, George C. 1961. *Human Behavior*. New York: Harcourt, Brace & World.

Horgan, John. 1996. *The End of Science*. Reading, MA: Addison-Wesley.

Hull, David L. 1988. *Science as a Process*. Chicago, IL: University of Chicago Press.

Hume, David. 1896 [1739]. *A Treatise in Human Nature*. Oxford: Clarendon Press.

Hunt, Alan. 1983. "Behavioral Sociology of Law: A Critique of Donald Black." *Journal of Law and Society*, 10: 19–46.

Kayama, M. and W. Haight. 2014. "Disability and Stigma: How Japanese Educators Help Parents Accept Their Children's Differences." *Social Work*, 59: 24–33.

Kitcher, Philip. 1982. *Abusing Science: The Case against Creationism*. Cambridge, MA: MIT Press.

Kitcher, Philip. 1985. *Vaulting Ambition*. Cambridge, MA: MIT Press.

Klinenberg, Eric. 2002. *Heat Wave: A Social Autopsy of Disaster in Chicago*. Chicago, IL: University of Chicago Press.

Kuhn, Thomas S. 1996 [1962]. *The Structure of Scientific Revolutions*, 3rd ed. Chicago, IL: University of Chicago Press.

Kuntz, Marcel. 2012. "The Postmodern Assault on Science: If All Truths Are Equal, Who Cares What Science Has to Say?" *EMBO Rep. 2012*, 10: 885–889.

Laquer, Walter. 1977. *Terrorism*. London: Weidenfeld and Nicolson.

Lessan, Gloria T. and Joseph F. Sheley. 1992. "Does Law Behave: A Macro Level Test of Black's Propositions on Change in Law." *Social Forces*, 70: 655–678.

Lindley, David. 1993. *The End of Physics*. New York: Basic Books.

Lindley, David. 2007. *Uncertainty: Einstein, Heisenberg, Bohr, and the Struggle for the Soul of Science*. New York: Anchor Books.

Lombardo, John A. 2008. *Three Generations, No Imbeciles: Eugenics, the Supreme Court, and Buck v. Bell*. Baltimore, MD: Johns Hopkins University Press.

Lopez, Robert S. 1976. *The Commercial Revolution of the Middle Ages, 950–1350*. Cambridge: Cambridge University Press.

Lopreato, Joseph. 1984. *Human Nature & Biocultural Evolution*. Boston, MA: Allen & Unwin.

Luper, Steven. 2004. "Epistemic Relativism." *Philosophical Issues*, 14: 271–295.

Lyman, Stanford. 1998. "Gunnar Myrdal's An American Dilemma after a Half Century: Critics and Anticritics." *International Journal of Politics, Culture and Society*, 12: 327–389.

Marshall, Douglas A. 2008. "The Dangers of Purity: On the Incompatibility of 'Pure Sociology' and Science." *Sociological Quarterly*, 49: 209–235.

Marx, Karl. 1909. *Capital*, Vol. 3. Chicago, IL: Kerr.

Marx, Karl. 1963 [1855]. *The Poverty of Philosophy*. New York: International Publishers.

Masterman, Margaret. 1970 [1965]. "The Nature of a Paradigm," pp. 59–89 in Imre Lakatos and Alan Musgrave, eds., *Criticism and the Growth of Knowledge*. London: Cambridge University Press.

Matson, Wallace. 2011. *Grand Theories and Everyday Beliefs*. New York: Oxford University Press.

Mayer-Schönberger, Viktor and Kenneth Cukier. 2013. *Big Data: A Revolution That Will Transform How We Live, Work and Think*. New York: Houghton Mifflin Harcourt Publishing.

McLanahan, Sara and Gary Sandefur. 1994. *Growing Up with a Single Parent: What Hurts, What Helps*. Cambridge, MA: Harvard University Press.

McRuer, Robert. 2006. *Crip Theory: Cultural Signs of Queerness and Disability*. New York: New York University Press.

Miller, Richard W. 1987. *Fact and Method*. Princeton, NJ: Princeton University Press.

Mish, Frederick C., ed. 2003. *Merriam-Webster's Collegiate Dictionary*, 11th ed. Springfield, MA: Meriam-Webster.

Moll, Jorge, Roland Zahn, Ricardo de Oliveira-Souza, Frank Krueger, and Jordan Grafman. 2005. "The Neural Basis of Human Moral Cognition." *Nature Reviews Neuroscience*, 6: 799–809.

Moore, Davis S. 2001. *The Dependent Gene*. New York: Henry Holt.

Morris, William, ed. 1969. *The American Heritage Dictionary of the English Language*. Boston, MA: American Heritage and Houghton Mifflin.

Moynihan, Daniel Patrick. 1965. *The Negro Family: The Case for National Action*. Washington, D.C.: United States Department of Labor.

Myrdal, Gunnar. 1944. *An American Dilemma: The Negro Problem and Modern Democracy*. New York: Harper.

Naroll, Raoul. 1967. "The Proposed HRAF Probability Sample." *Behavior Science Notes*, 2: 70–80.

Naroll, Raoul. 1968. "Some Thoughts on Comparative Method in Cultural Anthropology," pp. 248–253 in Hubert M. Blalock, Jr. and Ann B. Blalock, eds., *Methodology in Social Research*. New York: McGraw-Hill.

National Geographic. 2010. *Concise History of Science and Invention*, edited by Jolyon Goddard. Washington, D.C.: National Geographic.

Nietzsche, Friedrich W. 1968. *The Will to Power*, translated by Walter Kaufmann and R.J. Hollingdale. New York: Vintage.

Nosek, Brian. 2012. "An Open, Large-Scale, Collaborative Effort to Estimate the Reproducibility of Psychological Science." *Perspectives on Psychological Science*, 7: 657–660.

NOVA. 2003. "Viewpoints on String Theory." www.pbs.org/wgbh/nova/elegant/view-weinberg.html.

Omnes, Roland. 1999. *Quantum Philo*, translated by Arturo Sangalla. Princeton, NJ: Princeton University Press.

Orfield, Gary and Michal Kurlander, eds. 2001. *Diversity Challenge: Evidence on the Impact of Affirmative Action*. Cambridge, MA: Harvard Educational Publishing Group.

Otto, Shawn Lawrence. 2011. *Fool Me Twice: Fighting the Assault on Science in America*. Philadelphia, PA: Rodale Books.

Papineau, David. 1979. *Theory and Meaning*. Oxford: Clarendon Press.

Parsons, Keith, ed. 2003. *The Science Wars: Debating Scientific Knowledge and Technology*. Amherst, NY: Prometheus Books.

Parsons, Keith. 2014. *It Started with Copernicus*. Amherst, NY: Prometheus Books.

Parsons, Talcott. 1951. *The Social System*. Glencoe, IL: The Free Press.

Parsons, Talcott and Edward A. Shils, eds. 1951. *Toward a General Theory of Action*. New York: Harper & Row.

Pawelski, James G., Sarah Springer, J. Lane Tanner, Dennis L. Vickers, Ellen C. Perrin, Jane M. Foy, Carole E. Allen, James E. Crawford, Mark Del Monte, Miriam Kaufman, Jonathan D. Klein, and Karen Smith. 2006. "The Effects of Marriage, Civil Union, and Domestic Partnership Laws on the Health and Well-Being of Children." *Pediatrics*, 118: 349–364.

Pearl, Judea. 2009. *Causality: Models, Reasoning, and Inference*. New York: Cambridge University Press.

Phillips, Coretta. 2011. "Institutional Racism and Ethnic Inequalities: An Expanded Multilevel Framework." *Journal of Social Policy*, 40: 173–190.

Phillips, Julie A. 2013. "Factors Associated with Temporal and Spatial Patterns in Suicide Rates Across U.S. States, 1976-2000." *Demography*, 50: 591–614.

Pickett, Joseph P., ed. 2011. *The American Heritage Dictionary of the English Language*, 5th ed. Boston, MA: Houghton Mifflin Harcourt.

Piketty, Thomas. 2014. *Capital in the Twenty-First Century*. Cambridge, MA: Belknap Press of Harvard University Press.

Pinker, Steven. 2009. *How the Mind Works*. New York: W. W. Norton.

Popper, Karl. 1965. *The Logic of Scientific Discovery*. London: Hutchinson.

Poulantzas, Nicos. 1975. *Classes in Contemporary Capitalism*. London: New Left Books.

Powdermaker, Hortense. 1966. *Stranger and Friend: The Way of an Anthropologist*. New York: W. W. Norton.

Quine, W.V. 1951. "Two Dogmas of Empiricism." *Philosophical Review*, 60: 20–43.

Quine, W.V. 1960. *Word and Object*. Cambridge, MA: MIT Press.

Regnerus, Mark. 2012. "How Different are the Adult Children of Parents Who Have Same-Sex Relationships? Findings from the New Family Structures Study." *Social Science Research*, 41: 752–770.

Rigden, John S., ed. 1996. *Macmillan Encyclopedia of Physics*, Vol. 4. New York: Simon & Schuster Macmillan.

Rosanvallon, Pierre. 2013. *The Society of Equals*. Cambridge, MA: Harvard University Press.

Roscigno, Vincent J. 2011. "Power Revisited." *Social Forces*, 90: 349–374.

Ross, Andrew, ed. 1996. *Science Wars*. Durham, NC: Duke University Press.

Ryan, William. 1971. *Blaming the Victim*. New York: Random House.

Sanderson, Stephen K., Seth B. Abrutyn, and Kristopher R. Proctor. 2011. "Testing the Protestant Ethic Thesis with Quantitative Historic Data: A Research Note." *Social Forces*, 89: 905–911.

Schensul, Stephen L., Jean J. Schensul, and Margaret Diane LeCompte. 1999. *Essential Ethnographic Methods: Observations, Interviews, and Questionnaires*. Walnut Creek, CA: Altamira Press.

Schlefer, Jonathan. 2012. *The Assumptions That Economists Make*. Cambridge, MA: Harvard University Press.

Schütz, Alfred. 1967. *Phenomenology of the Social World*. Evanston, IL: Northwestern University Press.

Schwandt, Thomas A. 1996. "Farewell to Criteriology." *Qualitative Inquiry*, 2: 58–72.

Sewell, William H., Robert M. Hauser, Kristen W. Springer, and Taissa S. Hauser. 2001. "As We Age: A Review of the Wisconsin Longitudinal Study, 1957–2001." Madison, WI: CDE Working Paper No. 2001–2009. www.ssc.wisc.edu/cde/cdewp/2001-09.pdf.

Shattuck, Roger. 1996. *Forbidden Knowledge: From Prometheus to Pornography*. New York: St. Martin's Press.

Sheldrake, Rupert. 2012. *Science Set Free*. New York: Deepak Chopra Books.

Sherman, Lawrence. 1978. "Review of the Behavior of Law." *Contemporary Sociology*, 7: 11–15.

Shermer, Michael. 2005. "Science and Pseudoscience," pp. 669–673 in Donald M. Borchert, ed., *Encyclopedia of Philosophy, 2006*, 2nd ed., Vol. 8. Detroit, MI: Macmillan Reference, USA, an imprint of Thomas Gale.

Simon, Herbert A. 1952. "On the Definition of the Causal Relation." *Journal of Philosophy*, 49:517–528.

Simpson, J.A. and E.S.C. Weiner, eds. 1989. *The Oxford English Dictionary*, 2nd ed., Vols. 14 and 17. Oxford: Clarendon Press.

Sjoberg, Gideon and Roger Nett. 1968. *A Methodology for Social Research*. New York: Harper & Row.

Smolin, Lee. 2006. *The Trouble with Physics*. Boston, MA: Houghton Mifflin.

Smolin, Lee. 2013. *Time Reborn*. Boston, MA: Houghton Mifflin Harcourt.

Sober, E. 1999. "Testability." *Proceedings and Addresses of the American Philosophical Association*, 73: 47–76.

Sokal, Alan, and Jean Bricmont. 1998. *Fashionable Nonsense*. New York: Picador.

Southern, David. 1987. *Gunnar Myrdal and Black–White Relations: The Use and Abuse ofan American Dilemma, 1944–1969*. Baton Rouge: Louisiana State University Press.

Spradley, James P. 1970. *You Owe Yourself a Drunk: An Ethnography of Urban Nomads*. Boston, MA: Little, Brown.

Stack, Steven and Ain Haas. 1984. "The Effect of Unemployment Duration on National Suicide Rates: A Time Series Analysis, 1948–1982." *Sociological Focus*, 17: 17–29.

Stannard, Russell. 2010. *The End of Discovery*. Oxford: Oxford University Press.

Stark, Rodney. 2005. *The Victory of Reason*. New York: Random House.

Stevenson, Angus and Christine A. Linberg, eds. 2010. *New Oxford American Dictionary*, 3rd ed. New York: Oxford University Press.

Taylor, Charles L. and David A. Jodice. 1983. *The World Handbook of Political and Social Indicators*, 3rd ed., Vol. 2: *Political Protest and Government Change*. New Haven, CT: Yale University Press.

Tcherni, Maria. 2011. "Structural Determinants of Homicide: The Big Three." *Journal of Quantitative Criminology*, 27: 475–496.

Thagard, Paul. 2012. *The Cognitive Science of Science*. Cambridge, MA: MIT Press.

Thomas, W.I. and D.S. Thomas. 1928. *The Child in America*. New York: Knopf.

Thompson, Warren S. 1929. "Population." *American Journal of Sociology*, 34: 959–975.

Timmermans, Stefan and Iddo Tavory. 2012. "Theory Construction in Qualitative Research." *Sociological Theory*, 30: 167–186.

Tindall, D.B. 2000. "Some Misconceptions about Scientific Sociological Theory." *Sociological Focus*, 33: 57–77.

Tittle, Charles R. 1996. "Book Review." *Social Forces*, 75: 346–347.

Tooby, J. and Cosmides, L. 2005. "Conceptual Foundations of Evolutionary Psychology," pp. 5–67 in D.M. Buss, ed., *The Handbook of Evolutionary Psychology*. Hoboken, NJ: Wiley.

Trumble, Dennis R. 2013. *The Way of Science*. Amherst, NY: Prometheus Books.

Turner, Jonathan. 2010. *Theoretical Principles of Sociology*, Vol. 1: *Macrodynamics*. New York: Springer.

Uggen, Christopher, Jeff Manza, and Angela Behrens. 2004. "Less than the Average Citizen: Stigma, Role Transition and the Civic Reintegration of Convicted Felons," pp. 261–293 in Shadd Maruna and Ross Immarigeon, eds., *After Crime and Punishment: Pathways of Offender Reintegration*. Portland, OR: Willan Publishing.

Uhlmann, Allen J. 2004. "The Sociology of Subjectivity, and the Subjectivity of Sociologists: A Critique of the Sociology of Gender in the Australian Family." *British Journal of Sociology*, 55: 79–97.

Unzicker, Alexander. 2013. *Bankrupting Physics*. New York: Palgrave Macmillan.

van den Berghe, Pierre L. 1979. *Human Family Systems*. New York: Elsevier.

Wade, Nicholas. 2014. *A Troublesome Inheritance*. New York: Penguin Press.

Walker, Peter M.B., ed. 1999. *Chambers Dictionary of Science and Technology*. Edinburgh: Chambers Harrap.

Weber, Max. 1930. *Protestant Ethic and the Spirit of Capitalism*. London: Allen & Unwin.

Weinberg, Steven. 2009. *Lake Views*. Cambridge, MA: Harvard University Press.

Weinberg, Steven. 2015. *To Explain the World*. New York: Harper Collins Publishers.

White, Curtis. 2013. *The Science Delusion: Asking the Big Question in a Culture of Easy Answers*. Brooklyn, NY: Melville House.

Wilkinson, Richard G. and Kate E. Pickett. 2009. "Income Inequality and Social Dysfunction." *Annual Review of Sociology*, 35: 493–511.

Wilson, Edward O. 1978. *On Human Nature*. Cambridge, MA: Harvard University Press.

Witten, Ian H. and Eibe Frank. 2011. *Data Mining: Practical Machine Learning Tools and Techniques*, 3rd ed. Burlington, MA: Elsevier.

Woit, Peter. 2006. *Not Even Wrong*. New York: Basic Books.

Wong, Kam C. 1998. "Black's Theory on the Behavior of Law Revisited III: Law as More or Less Governmental Social Control." *International Journal of the Sociology of Law*, 26: 365–392.

Wong, Raymond S. 1990. "Understanding Cross-National Variation in Occupational Mobility." *American Sociological Review*, 55: 560–573.

Woodward, C. Vann. 1955. *The Strange Career of Jim Crow*. New York: Oxford University Press.

Wootton, David. 2015. *The Invention of Science*. New York: Harper.

Wright, Erik O. 1978. *Class, Crisis and the State*. London: New Left Books.

Wright, Erik O. 1979. *Class Structure and Income Determination*. New York: Academic Press.

Wright, Erik O., Cynthia Costello, David Hachen, and Joey Sprague. 1982. "The American Class Structure." *American Sociological Review*, 47: 709–726.

Wuthnow, Robert. 2014. *Rough Country: How Texas Became America's Powerful Bible-Belt State*. Princeton, NJ: Princeton University Press.

Yip, Paul S.F. and Eric D. Caine. 2011. "Employment Status and Suicide." *Journal of Epidemiology and Community Health*, 65: 733–736.

York, Richard and Bret Clark. 2006. "Marxism, Positivism, and Scientific Sociology." *Sociological Quarterly*, 47: 425–450.

Zaki, Mohammed J. and Wagner Meira, Jr. 2014. *Data Mining and Analysis: Fundamental Concepts and Algorithms*. New York: Cambridge University Press.

Zimbardo, Philip G. 1972. "Pathology of Imprisonment." *Society*, 9: 4–8.

Zimbardo, Philip G. 2007. *The Lucifer Effect: Understanding How Good People Turn Evil*. New York: Random House.

NAME INDEX

SUBJECT INDEX